新楼盘 39

NEWHOUSE 图解地产与设计

旅游度假区

中国林业出版社

深圳大学基础实验楼二期

深圳中海圣廷苑酒店

深圳大鹏地质博物馆

深圳莱蒙水榭春天

深圳星河时代花园

奥德景观
LUCAS DESIGN GROUP

深圳市奥德景观规划设计有限公司简介

公司坐落于著名的蛇口湾畔，深圳最有影响力的创意设计基地：南海意库；

公司前身为深圳市卢卡斯景观设计有限公司，是由 2003 年成立于香港的卢卡斯联盟（香港）国际设计有限公司在世界设计之都：深圳设立的中国境内唯一公司。

于 2012 年 1 月获得中华人民共和国国家旅游局正式认定：旅游规划设计乙级设计资质。

公司专注于：

居住区景观与规划设计（含旅游地产）
商业综合体景观与规划设计（含购物公园、写字楼及创意园区）
城市规划及空间设计
市民公园设计
酒店与渡假村景观规划与设计
旅游策划及规划设计

公司目标：倚重当下的中国的文化渊源结合世界潮流，尊重地域情感，在中国打造具强烈地域特征的、风格化的、国际化的，具前瞻性、可再生的的城市景观、人居环境、风情渡假区及自然保护区。

地　址：深圳市南山区蛇口兴华路南海意库 2 栋 410 室
电　话：0755-86270761
传　真：0755-86270762
邮　箱：lucasgroup_lucas@163.com
网　址：www.lucas-designgroup.com

南宁中铁凤岭山语城

惠州御湾雅墅

西西安小镇

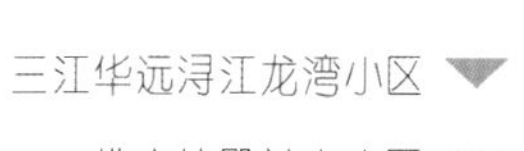

三江华远浔江龙湾小区

佛山桂殿兰宫小区

山东烹饪学院

保定燕赵国际城市综合体

毕节大众国际花园

南宁太阳岛小区

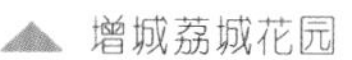

增城荔城花园

王志纲丽江书院

常德西城中环城市综合体

梅州山水城项目

柳州山水一号项目

Web: http://www.kingmade.com

Tel:020-88832190 88832191

Add:广州市天河区珠江西路5号

广州国际金融中心主塔写字楼第21层06-08单元

深圳市佰邦建筑设计顾问有限公司 作为国内具有影响力的建筑设计公司之一，在中国深圳、福州、香港均设有分支。专业从事大型社区的规划和城市设计，大型公共建筑的设计，涉及城市综合体，文化教育建筑、会展建筑、商业建筑、酒店和交通建筑等。

佰邦建筑一直致力建立协同设计机制，联合对等的相关地产服务企业，组织各自独立、运行高效的设计联合体。

除了有创新的设计理念，我们还有一支朝气蓬勃的员工队伍，以优秀的工作绩效和工作热情实现创造着佰邦的一切。

前言 EDITOR'S NOTE

社会、生态与功能的三位一体

Trinity of Society, Ecology and Function

旅游度假是人们度过闲暇时间的一种重要方式，随着经济的发展、科技的进步，人们自由支配的金钱和时间越来越多，旅游度假的需求也随之不断上升。

旅游度假区作为旅游目的地类型之一，并不完全等同于传统意义上的“观光”游览区。它通过提供娱乐设施和服务项目创造一种愉快欢乐，或是宁静舒适的环境，力图达到旅游者放松身心的目的。因此，旅游度假区的规划设计，就是要创造出一种健康舒适、功能完备、生态和谐的旅游度假空间。规划中不仅要将周边的著名观光景点纳入度假区范围内，更要设置充足的基础服务设施，满足旅游度假者的食宿、康体、娱乐、休闲、会议等多种需求。

随着旅游业的高速发展，旅游度假将成为人们一种必不可少的新的生活方式。而我国的旅游度假区兴起较晚，因此，深入地了解旅游度假区的发展现状，客观分析其未来的发展趋势，对我国旅游度假区的建设将有着非常重要的作用。本期旅游度假区专题集结了一些特色旅游度假区项目，通过丰富的技术图纸、文字对其展开详细、深入的剖析，希望能为读者带来不一样的启发。

Travel is an important way of life in spare time. With the development of the economy and the advance of science and technology, we have more and more time and money for travel.

As one of the destinations for travel, the tourism resort is quite different from the traditional "sightseeing" areas. Equipped with entertainment facilities and services, it will provide an environment for happiness and pleasure, or for peace and comfort, relaxing the tourists to the utmost. Therefore, the tourism resort planning needs to create a tourism space which is healthy and comfortable, well-equipped and functional, ecological and harmonious. It will not only bring in the surrounding scenic spots but also set adequate service facilities to meet the requirements for accommodation, health, entertainment, recreation, conference, etc.

With the rapid development of the tourism industry, travel for holiday will become an essential part of our life. While China starts later in this industry, so deep research to the current situations and objective analysis on the development trend will make a great difference to the construction of the tourism areas. In this issue, we have selected some typical cases in tourism planning and design, and done in-depth analysis with full drawings, images and text description. Hope it brings new inspirations to the readers.

王志

jiatu@foxmail.com

NEWHOUSE 图解地产与设计

2012年 总第39期

面向全国上万家地产商决策层、设计院、建筑商、材料商、专业服务商的精准发行

图书在版编目（CIP）数据

新楼盘. 旅游度假区：汉英对照 / 佳图文化主编.
-- 北京：中国林业出版社, 2012.7
ISBN 978-7-5038-6689-0

Ⅰ. ①新… Ⅱ. ①佳… Ⅲ. ①建筑设计 - 中国 - 现代 - 图集 Ⅳ. ①TU206

中国版本图书馆CIP数据核字(2012)第094523号
出版：中国林业出版社
主编：佳图文化
责任编辑：李顺 许琳
印刷：利丰雅高印刷(深圳)有限公司

编辑部地址: 广州市海珠区新港西路3号银华大厦4楼
电话: 020-89090386/42/49、28905912
传真: 020-89091650

北京办: 王府井大街277号好友写字楼2416
电话: 010-65266908 **传真:** 010-65266908

深圳办: 深圳市福田区彩田路彩福大厦B座23F
电话: 0755-83592526 **传真:** 0755-83592536

副理事长单位 DEPUTY CHAIRMAN

华森建筑与工程设计顾问有限公司　邓明 广州公司总经理
地址：深圳市南山区滨海之窗办公楼6层
广州市越秀区德政北路538号达信大厦26楼
TEL：0755-86126888　020-83276688
http://www.huasen.com.cn　E-mail:hsgzaa@21cn.net

上海中建建筑设计院有限公司　徐峰 董事长
地址：上海市浦东新区东方路989号中达广场12楼
TEL：021-68758810　FAX：021-68758813
http://www.shzjy.com
E-mail：csaa@shzjy.com

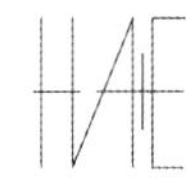

广州瀚华建筑设计有限公司　冼剑雄 董事长
地址：广州市天河区黄埔大道中311号羊城创意产业园2-21栋
TEL：020-38031268　FAX：020-38031269
http://www.hanhua.cn
E-mail：hanhua-design@21cn.net

常务理事单位 EXECUTIVE DIRECTOR OF UNIT

深圳市华域普风设计有限公司　梅坚 执行董事
地址：深圳市南山区海德三道海岸城东座1306-1310
TEL：0755-86290985　FAX：0755-86290409
http://www.pofart.com

天萌（中国）建筑设计机构　陈宏良 总建筑师
地址：广州市天河区员村四横路128号红专厂F9栋天萌建筑馆
TEL：020-37857429　FAX：020-37857590
http://www.teamer-arch.com

天友建筑设计股份有限公司　马素明 总建筑师
地址：北京市海淀区西四环北路158号慧科大厦7F（方案中心）
TEL：010-88592005　FAX：010-88229435
http://www.tenio.com

奥雅设计集团　李宝章 首席设计师
深圳总部地址：深圳蛇口南海意库5栋302
TEL：0755-26826690　FAX：0755-26826694
http://www.aoya-hk.com

广州山水比德景观设计有限公司　孙虎 董事总经理兼首席设计师
地址：广州市天河区珠江新城临江大道685号红专厂F19
TEL：020-37039822/823/825　FAX：020-37039770
http://www.gz-spi.com

广州市四季园林设计工程有限公司　原帅让 总经理兼设计总监
地址：广州市天河区龙怡路117号银汇大厦2505
TEL：020-38273170　FAX：020-86682658
http://www.gz-siji.com

深圳市佰邦建筑设计顾问有限公司　迟春儒 总经理
地址：深圳市南山区兴工路8号美年广场1栋804
TEL：0755-86229594　FAX：0755-86229559
http://www.pba-arch.com

北京博地澜屋建筑规划设计有限公司　曹一勇 总设计师
地址：北京市海淀区中关村南大街31号神舟大厦8层
TEL：010-68118690　FAX：010-68118691
http://www.buildinglife.com.cn

香港华艺设计顾问（深圳）有限公司　林毅 总建筑师
地址：深圳市福田区华富路航都大厦14、15楼
TEL：0755-83790262　FAX：0755-83790289
http://www.huayidesign.com

华通设计顾问工程有限公司
地址：北京市西城区西直门南小街135号西派国际C-Park3号楼
TEL：8610-83957395　FAX：8610-83957390
http://www.wdce.com.cn

GVL国际怡境景观设计有限公司　彭涛 中国区董事及设计总监
地址：广州市珠江新城华夏路49号津滨腾越大厦南塔8楼
TEL：020-87690558　FAX：020-87697706
http://www.greenview.com.cn

R-LAND 北京源树景观规划设计事务所　白祖华 所长
地址：北京朝阳区朝外大街怡景园5-9B
TEL：010-85626992/3　FAX：010-85625520
http://www.ys-chn.com

北京寰亚国际建筑设计有限公司　赵士超 董事长
地址：北京市朝阳区琨莎中心1号楼1701
TEL：010-65797775　FAX：010-84682075
http://www.hygjjz.com

奥森国际景观规划设计有限公司　李焯忠 董事长
地址：深圳市南山区南海大道粤海路动漫园7栋5楼
TEL：0755-26828246 86275795　FAX：0755-26822543
http://www.oc-la.com

深圳市雅蓝图景观工程设计有限公司　周敏 设计董事
地址：深圳市南山区南海大道2009号新能源大厦A座6D
TEL：0755-26650631/26650632　FAX：0755-26650623
http://www.yalantu.com

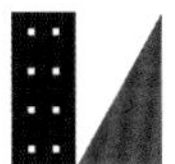

北京新纪元建筑工程设计有限公司　曾繁柏 董事长
地址：北京市海淀区小马厂6号华天大厦20层
TEL：010-63483388　FAX：010-63265003
http://www.bjxinjiyuan.com

HPA上海海波建筑设计事务所　陈立波、吴海青 公司合伙人
地址：上海中山西路1279弄6号楼国峰科技大厦11层
TEL：021-51168290　FAX：021-51168240
http://www.hpa.cn

哲思（广州）建筑设计咨询有限公司　郑竞晖 总经理
地址：广州市天河区天河北路626号保利中宇广场A栋1001
TEL：020-38823593　FAX：020-38823598
http://www.zenx.com.au

理事单位 COUNCIL MEMBERS (排名不分先后)

广州市柏澳景观设计有限公司　徐农思 总经理
地址：广州市天河区龙怡路91号农机物资公司综合楼东梯六楼
TEL：020-87569202
http://www.bacdesign.com.cn

北京奥思得建筑设计有限公司　杨承冈 董事总经理
地址：北京朝阳区东三环中路39号建外SOHO16号楼2903~2905
TEL：86-10-58692509/19/39　FAX：86-10-58692523

广州嘉柯园林景观设计有限公司　陈航 执行董事
地址：广州市珠江新城华夏路49号津滨腾越大厦北塔506-507座
TEL：020-38032521/23　FAX：020-38032679
http://www.jacc-hk.com

CDG国际设计机构　林世彤 董事长
地址：北京海淀区长春路11号万柳亿城中心A座10/13层
TEL：010-58815603 58815633　FAX：010-58815637
http://www.cdgcanada.com

上海唯美景观设计工程有限公司　朱黎青 董事、总经理
地址：上海徐虹中路20号2-202室
TEL：021-61122209　FAX：021-61139033
http://www.wemechina.com

深圳灵顿建筑景观设计有限公司　刘海东 董事长
地址：深圳福田区红荔路花卉世界313号
TEL：0755-86210770　FAX：0755-86210772
http://www.szld2000.com

中房集团建筑设计有限公司　范强 总经理 / 总建筑师
地址：北京市海淀区百万庄建设部院内
TEL：010-68347818

陈世民建筑师事务所有限公司　陈世民 董事长
地址：深圳市福田中心区益田路4068号卓越时代广场4楼
TEL：0755-88262516/429

侨恩国际（美国）建筑设计咨询有限公司
地址：重庆市渝北区龙湖MOCO4栋20-5
TEL：023-88197325　FAX：023-88197323
http://www.jnc-china.com

广州市圆美环境艺术设计有限公司　陈英梅 设计总监
地址：广州市海珠区宝岗大道杏坛大街56号二层之五
TEL：020-23353942　FAX：020-34267226
http://www.gzyuanmei.com

上海金创源建筑设计事务所有限公司　王毅 总建筑师
地址：上海杨浦区黄兴路1858号701-703室
TEL：021-55062106　FAX：021-55062106-807
http://www.odci.com.cn

深圳市奥德景观规划设计有限公司　凌敏 董事总经理、首席设计师
地址：深圳市南山区蛇口海上世界南海意库2栋410#
TEL：0755-86270761　FAX：0755-86270762
http://www.lucas-designgroup.com

目录

CONTENTS

068

110

新特色 NEW CHARACTERISTICS

新空间 NEW SPACE

新创意 NEW IDEA

124

商业地产 COMMERCIAL BUILDINGS

134

“刚改”需求未来或成楼市供需的主流

随着越来越多的中高端住宅开始寻求入市，被称为“刚改”的改善需求或将接下“刚需”接力棒，主导市场继续上行。北京市房协秘书长陈志等业内人士表示，改善性需求稳步释放，未来一段时间内有可能将成为楼市供需的主流，这部分需求除了开发商的“以价换量”外，更加看重项目的质量和性价比。

Reasonable Improvement Demand About to be the Mainstream of Supply and Demand in the Property Market

As more and more middle and high-end residential seek to enter the market, reasonable improvement demand is about to take over the rigid demand to dominate the market. Chen Zhi, secretary-general of Beijing Housing Association, expresses that reasonable improvement demand will become the mainstream of supply and demand in the property market in the next period of time little by little. In addition to the "price change" of the developers, this part of demand pays much more attention to the quality and cost-effect of the projects.

中弘股份携手万豪主攻旅游地产

在住宅地产进入冬天之际，中弘股份将视线瞄准了旅游地产。6月8日，中弘携旗下位于海口和北京的两家酒店，与世界知名酒店管理公司万豪国际集团进行了管理合作项目的签约。“此次签约旨在展开双方合作的序幕，未来双方会在发展布局、市场协同、品牌联合等方面有着纵深的合作，实现优势互补。”中弘股份董事长王永红在签约现场如是表示。

Zhonghong and Marriott March to Tourist Real Estate Hand in Hand

Zhonghong aims at tourist real estate when residential real estate tipped into recession. On June 8th, together with its two hotels in Haikou and Beijing, Zhonghong signed a contract with the world-renowned hotel management company Marriott. "This contract is singed to begin the bilateral cooperation in future development, market synergy and co-branding. "said the chairman of Zhonghong Wang Yonghong when signing the contract.

北京公租房将借鉴“香港模式”

近日举行的京港洽谈会住房保障工作座谈会上，香港的公屋管理模式引发了北京市相关部门的兴趣。市住保办相关负责人说，12月1日起全市街道乡镇公租房申请窗口即将打开，在随后就会产生的公租房管理问题，本市将借鉴香港等城市成熟的管理经验，对公租房的配租、管理精细化。

Beijing Public Rental Will Imitate "Hong Kong Pattern"

On the recent Beijing-Hong Kong Fair Housing Security Forum, Hong Kong's public housing management pattern has sparked the interest of the relevant departments in Beijing. Person in charge of the housing security office said that public rental application windows in all towns and villages will open on December 1st, the subsequent rental management issues will be coped with the mature management experience of Hong Kong.

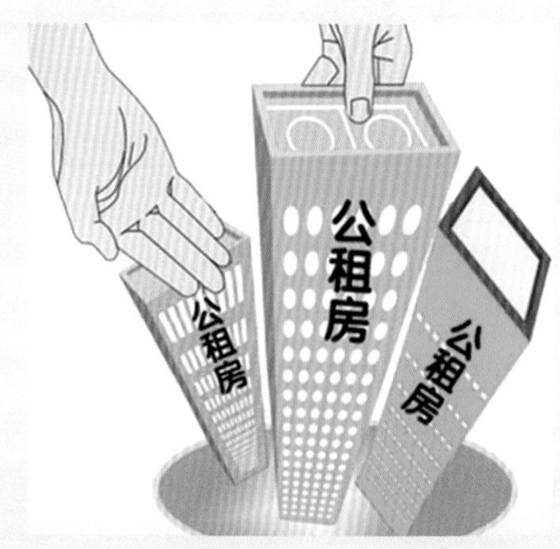

国土新规再治囤地

6月7日，国土资源部发布中华人民共和国国土资源部令，《闲置土地处置办法》自2012年7月1日起施行。文件中“闲置土地满一年按地价款20%征缴土地闲置费，闲置两年则无偿收回。”为最大的“亮点”。

New Regulations Again to Manage Land Reserve

On June 7, the Ministry of Land and Resources issued an order, "Idle Land Disposal Methods", which will be executed from July 1, 2012. The land idle for one year will be taxed inactivity fees that of 20% of the land price, idle for two years will be taken back with no pay" is the largest bright spot.

央行：6月8日起降息0.25个百分点

6月7日晚间，中国人民银行决定，自2012年6月8日起下调金融机构人民币存贷款基准利率。其中，金融机构一年期存款基准利率下调0.25个百分点，一年期贷款基准利率下调0.25个百分点；其他各档次存贷款基准利率及个人住房公积金存贷款利率相应调整。

Central Bank: Cut Interest Rates by 0.25 Percentage Points from June 8

On the evening of June 7, the People's Bank of China decided to cut financial institutions RMB benchmark deposit and lending interest rates since June 8, 2012. Among them, the one-year benchmark deposit rate cut by 0.25 percentage points and so did the lending interest rate. Other deposit and lending interest rate in different levels and that of individual housing provident fund will be adjusted accordingly.

银监会：二套房贷将与首套房贷“一视同仁”

中国版巴塞尔III终于露出真容。近日，银监会正式发布《商业银行资本管理办法(试行)》，并于2013年1月1日起实施。值得注意的是，正式文件对二套房贷与首套房贷给予同样的风险权重。业内人士认为，这一调整有效降低了二套房贷的资金成本，可能会促进银行发放二套房贷款。

CBRC: Mortgage Loan of the Second Home will be Equal to That of the First Suite

Chinese version of Basel III is finally unraveling. Recently, the CBRC officially released commercial bank's capital management (Trial) will be implemented from January 1, 2013. It highlights that the official document treat Mortgage loans for the second home and the first suite equally. The industry believes that this adjustment effectively reduces the cost of the second home mortgage and may contribute to Bank issued loads for the second home.

楼市限购与稳增长如何端平

近期，国务院常务会议指出目前经济下行压力加大，要求把稳增长放在更加重要的位置。在“增长”的诸项措施中提到，要稳定和严格实施房地产市场调控政策，同时加大对保障房的支持力度。那么，如何在执行房地产调控的同时，实现“稳增长”呢？多位专家认为，“稳增长”要坚持限购控制房价，同时不断满足刚性购房者的需求。

How to Balance House Purchase Restriction and Steady Growth

Recently, the State Council executive meeting points out that steady growth should be placed in a more important position due to the increasing economic downward pressure. Mentioned in the various measure of growth, real estate market regulation policies should be implemented strictly while increasing support to indemnificatory housing. Then, how to achieve steady growth while implementing the restricted order? Many experts believe that steady growth has to meet the needs of rigid buyers when adhering to restriction and price-controlling.

“抱团取暖——探索民营建筑设计公司寒冬举措”论坛

5月26日上午，在庄重的中华圣公会教堂，由北京工程勘察设计行业协会民营工作委员会、新纪元建筑设计公司设计主办，华太设计、墨臣设计协办的“抱团取暖——探索民营建筑设计公司的寒冬举措”研讨会举办。

北京工程勘察设计行业协会秘书长周荫如发表致辞，北京市规划委员会委员毛哲、北京工程勘察设计行业协会会长曲际水出席了论坛。而来自CCDI悉地国际、新纪元设计、墨臣设计、九源国际、华太设计、维拓时代等众多优秀的行业企业家，针对市场转向下的民营建筑设计企业未来发展方向、公司管理创新、人力培训、品牌建设、行业组织协调等热点问题，分别在论坛发表主题演讲，并与相关行业领导及行业内的总建筑师进行了热烈讨论。同时，2012年，北京新纪元建筑工程设计有限公司作为年度北京工程勘察设计行业协会民营工作委员会轮值主席单位，在论坛上也举行了民营协会轮值主席单位的交接仪式。

论坛也得到业内业外众多如《安家》、《中国房地产报》、《UED》、筑龙网、《京华时报》、《新京报》等主流媒体的支持。

Forum on Cooperation—Private Architecture Firms Explore Measures to Pull Through

In the morning of May 26th, 2012, Forum on Cooperation—Private Architecture Firms Explore Measures to Pull Through was held in the solemn Mochen Holy Catholic Church of China, which sponsored by Beijing Engineering Survey & Design Industry Association Private Work Committee and Beijing New Era Architectural Design Ltd., and presented by SINO-SUN and MoChen.

Zhou Yinru, the secretary-general of Beijing Engineering Survey & Design Industry Association, delivered a speech. Mao Zhe, a member of Beijing Municipal Commission Association, and Qu Jishui, the chairman of Beijing Engineering Survey & Design Industry Association attended the forum. Numerous outstanding entrepreneurs from CCDI, New Era, MoChen, Jiuyuan Group, SINO-SUN and Victory Star issued speeches respectively on the future development, management innovation, HR training, brand building, organization and coordination of private architecture firms during the turning of the market, and held heated discussions with the leadership in relevant industries and chief architects in this industry as well. Besides, as the new chairman, Beijing New Era Architectural Design Ltd. held the private association rotating chairman handover ceremony.

The forum got the support of the mainstream media, such as ihome, China Real Estate Business, UED, Zhu Long, Beijing Times and Beijing News.

巴拉腊特艺术画廊附属建筑

项目的建造涵盖不同的形式，包括公共大厅、凉亭、乐队演奏台，这些因素都在形式上或是功能上反映了当地的建造形式。项目将一个简单的活动空间转化为一个灵活的回应场地及项目需求的设施，为人们提供多功能的演讲、工作室、装置及画廊等功能空间。

Annexe for the Art Gallery of Ballarat

The project shifts across hybrid postures of public hall, verandah and bandstand, formally and programmatically recalling local typologies. The project transforms a simple "events marquee" brief into a flexible response to site and program providing an enclosed multi-purpose space for talks, workshops, installations and gallery functions.

天津美术馆

项目由干净的线条和自然的光线定义。这个美术馆将容纳四个永久展览，这四个部分包括中国书法、西方艺术、雕塑，和现代艺术，除此之外，馆内还包括其他临时性的展览空间。还有其他的三个文化设施：图书馆、歌剧院和另外一个新博物馆。

Tianjin Art Museum

The Tianjin Art Museum has space to house four permanent exhibitions. In addition to rooms for Chinese calligraphy, western art, sculpture, and modern art, there are also galleries in which changing exhibitions can be presented. There are three additional cultural facilities: a library, an opera house and another new museum.

海滨拾荒者博物馆

根据评委的意见，海滨拾荒者博物馆因为其透明性，几乎可以感受到室外任何气候的变化。“太阳、云彩、雷电和雨水，室外的变成了室内的，形成一种独特的观察感和情绪感，这就是这个位于瓦登海边上的建筑的核心质量。”

Maritime and Beachcombers Museum

One can almost feel the weather because of the transparency of the building, according to the jury. "Sun, clouds, thunder and rain: outdoors comes inside as perception and emotion and this is a core quality for a building with the Wadden Sea at your doorstep."

澳大利亚国立美术馆

澳大利亚国立美术馆及其周围的雕塑公园完工于1983年，相关机构对其进行翻新，包括新建一处大楼入口，扩建画廊，还有在南面新建一个公园。2005年，事务所受委托对美术馆的公共区域细节进行设计，新建的公园里有艺术感极强的雕塑。

National Gallery of Australia

First completed in 1983, the National Gallery of Australia stands itself as a work of art and a pillar of architectural splendor. Expansion and renovation consist of a new entrance, a gallery and a new garden proceeded in 2005. The newly-created Australian Garden includes a skyspace sculpture by American artist James Turrell.

丹德农政府服务办公室

这是由HASSELL事务所设计的丹德农政府服务办公室项目，这个项目是复兴丹德农中心地区总体规划项目的重要部分，设计的挑战是创造出高标准的城市设计，工作环境，从而激发综合利用的活力。

Dandenong Government Services Offices

The Government Services Offices development is one of the key projects of the Revitalizing Central Dandenong initiative, a major urban renewal project being designed by HASSELL. The challenge of this landmark development was to create a high standard of urban design and quality workplace in this outer-suburban region, and to assist in the rebirth of the precinct as a major mixed-use activity centre.

金色工作室

这是一个位于明斯特大学内部的临时展馆，非对称的十字形排布策略性地立于广场之上，从而从四个不同的分支结构中吸收外部的视野。黑色线条的窗户框定狭窄的门廊，它慢慢将人们引向内部的空间，同时突出了室外的行人，最终形成一种温暖的室内环境。

Golden Workshop

Golden Workshop is a temporary pavilion installed at the University of Mü nster, the asymmetrical cross-form plan is strategically located within the plaza to draw inward-looking views from each of its four wings. The black-outlined windows frame the narrowing corridors, tapering the interior walls towards the center, funneling inside the attention of passersby, generating a warm interior atmosphere.

JOH 3公寓楼

这是由J. Mayer H. Architects事务所设计的JOH 3公寓楼项目，它呈现出独特的表皮形式，重新诠释了经典的柏林住宅建筑形式，同时融入多单元的结构和绿色的内部庭院。悬挑板条的雕塑感设计来源于城市的景观特质，它是对城市轮廓和布局的一种影射。

JOH 3 Apartment

J. Mayer H. architects' design for the JOH 3 Apartment reinterprets the classic Berliner residential building with its multi-unit structure and green interior courtyard. The sculptural design of the suspended slat facade draws on the notion of landscape in the city, a quality visible in the graduated courtyard garden and the building's silhouette and layout.

Maree Basse住宅

项目主要围绕两个错离中心的体量展开，因而获得了最大的光照，同时也吸收了很多不同的视野。这两个体量看起来好似从森林中漂浮而生，向外突出，鸟瞰河流。整体住宅形成一种别致的角度，拥抱和遮盖巨大的位于低层的一个私人阳台。

Residence Mar é e Basse

The concept of the residence revolves around two off-centered volumes positioned in such a way to take in as much sunlight as possible and to make the most out of the different viewpoints. These two volumes seem to emerge from the forest and project themselves towards the River and La Malbaie. The whole house forms an angle that embraces a huge private terrace at the lower level.

津屋崎町住宅

这是一个住宅和艺术工作室项目。表皮包含垂直的部分，同时包含双曲线，它开始于中间，继而向内向外弯曲。一系列扭曲的元素延展于建筑的整个长度，它们在由统一的变形导致的空洞中形成窗户结构，并受到屋顶线条的保护。

House of Coast Work Tsuyazaki

This is a residence and artist's studio. The facade is comprised of vertical segments with a double curvature starting at the midpoint to warp inward and outward. Spanning the length of the building, the series of twisting elements form windows within the voids generated by the uniform deformations and are protected beneath the roofline.

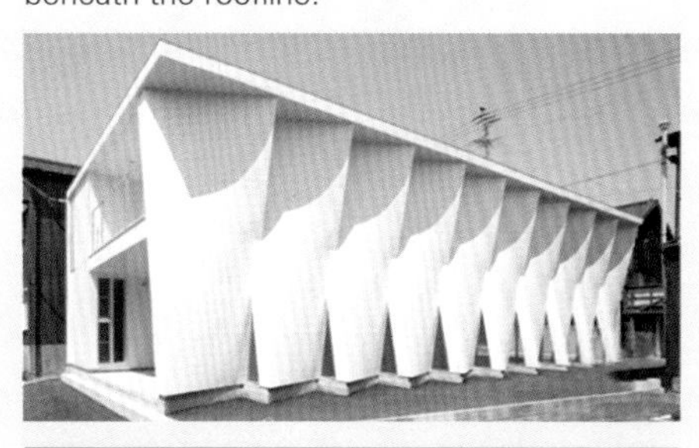

A-G住宅

这个项目重新理解经典的中世纪加州现代设计原则，并将其适用于澳大利亚传统住宅的设计当中，从而创造出现代的住宅模式。三个翼结构交织在一起，形成一个功能性的脊梁，同时典型的澳洲后院被重新调整了位置和朝向，变为了住宅的前侧院落。

A-G House

The A-G House reinterprets classic mid-century California Modern design principles to suit the Australian design vernacular, creating a contemporary home for a young at heart - three generational family. Three wings are connected via a functional spine, whilst the archetypal Australian backyard is reoriented to the front of the house in the form of a central, alfresco entertaining area that merges seamlessly with the indoor living space.

乡村农仓住宅

设计重新诠释农仓结构，室外由一层水平的木质遮挡定义，它使空间在进入到室内之前能够流通到一个暂时的过渡的覆盖空间之中，巨大的农仓式的门结构隐藏在覆盖的板面之下，它们可以关闭以保存隐私，或是打开以增加日光的吸收。

A Barn in the Countryside

Reinterpreting the barn as a home, the exterior is sheathed with a layer of horizontal wooden blinds, allowing air to ventilate into a transitory covered space before entering the interior. Concealed and integrated into the cladding, large barn-style doors may remain closed for privacy or swung open for additional daylight.

H2住宅

这是由314 Architecture Studio工作室设计的位于希腊雅典的住宅项目，它包含三个单层的公寓，每个250 m^2。每个公寓都有两个小的卧室和一个大卧室。在室内有一个巨大的中庭，它能为次级空间提供光线，同时也能充当一个漏斗结构，从而在夏季导出热气流。

H2 House

This building designed by 314 Architecture Studio is located at the suburbs of Athens in the area next to the golf court in Glyfada, and consists of three one floor apartments of 250 m^2 each. Each apartment has two small and one master bedroom. Inside the building there is an atrium that provides light to secondary areas of the apartments while also working as a funnel exit for warm air energy consumption in the summer months.

HOUSING WITH USER GENERATED CONTENT | Hoogvliet Social Housing, Rotterdam

为人度身订造的独特住宅—— 霍赫弗列特联排住宅

项目地点：荷兰鹿特丹霍赫弗列特
客　　户：Vestia
建筑设计：BBVH建筑设计事务所

Location: Hoogvliet, Rotterdam
Client: Vestia
Architectural Design: BBVH Architecten BV

项目概况

荷兰BBVH建筑师曾经不断地寻求一个串行住房项目解决方案，以使购房者能够对他们房屋的布局和大小产生更大的影响。BBVH并不仅仅想为2.31人的普通家庭设计房屋，而是要设计具有适合性的房屋。该项目正是在这样的理念设计下诞生的作品，项目位于荷兰鹿特丹霍赫弗列特，包含了30栋的联排住宅。

建筑设计

只因见过太多的、有着相同的布局、对于大家庭嫌太窄、对于无子女的夫妻又认为太宽敞的楼面布置图，设计师提出了不同的解决方案。

第一步：设计一座模块化的房子，可以对其轻易地进行改变。例如可以增加一个房间，增加一个楼层，改变布局，扩大起居室等等，所有这些变化都可以使用相同的基本模块来实现。这样，就可以在同一个联排中建造一栋135 m^2的、具有成本效益的小房子和一栋近200 m^2、有五间卧室的巨大的别墅式房屋。

第二步：建立一个在线配置器，使潜在购房者能够创建自己梦想中的房子。该软件一方面必须门槛很低且易于掌握，另一方面必须包含关于房子的所有相关的和重要的实际情况，如对各个项目的评价、面积、包含所做选择的建筑平面图等。这样形成了一个分步式的过程，人们可以逐步地设计自己的房子。

灵活的户型选择是设计的主题。通过使用交互式软件，潜在客户可以自由组合出他们的理想家园。对于基本的住宅项目，其楼层数和户型都是可以自由选择的。这样既形成了多样化的立面效果，同时同类材料和涂层的使用也保证了整体的统一性。所有的住宅均拥有两种主立面、超高的客厅和理想的朝向，尽享壮观的运河景观。

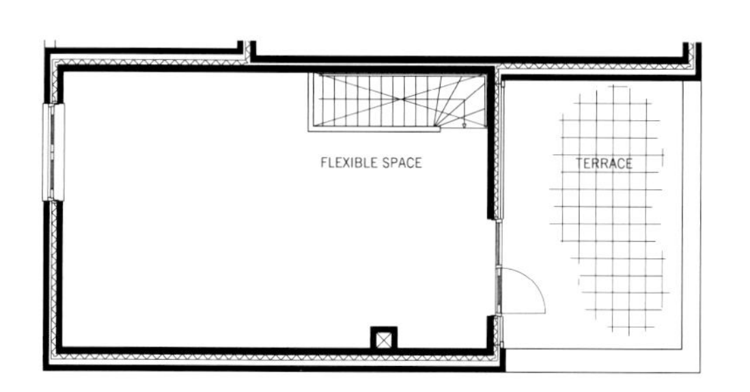

FLOORPLAN LEVEL 2 - OPTIONAL

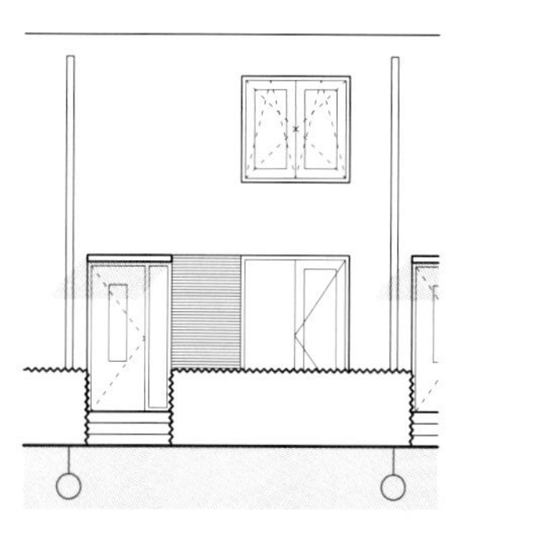

FRONT FACADE

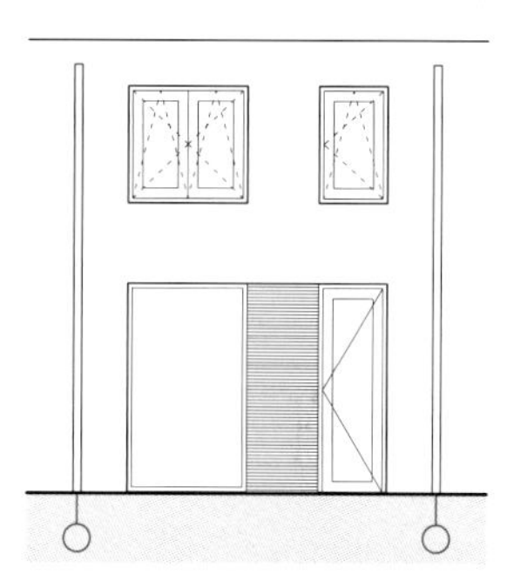

BACK FACADE

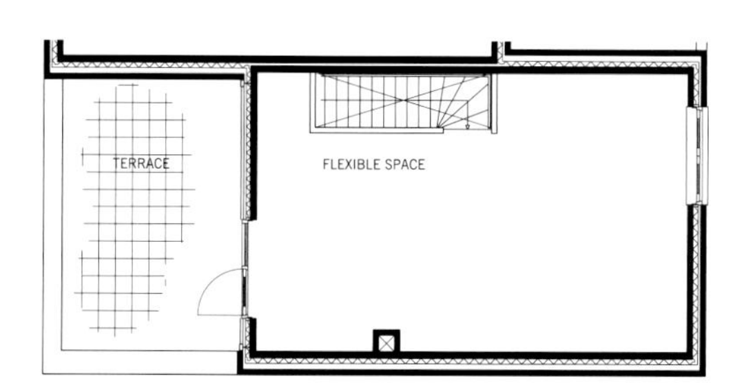

FLOORPLAN LEVEL 2 - OPTIONAL

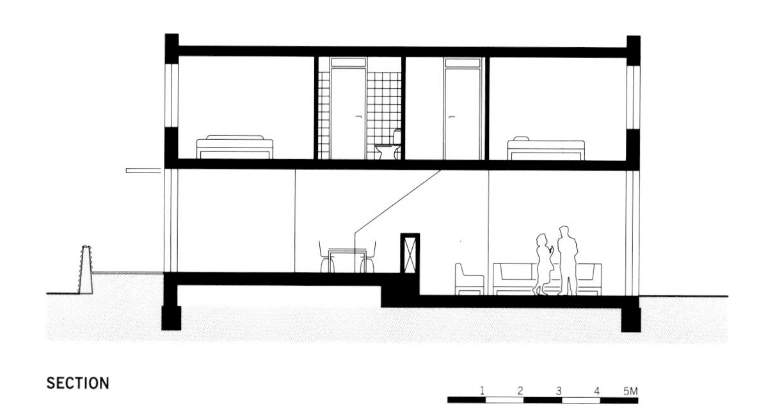

SECTION

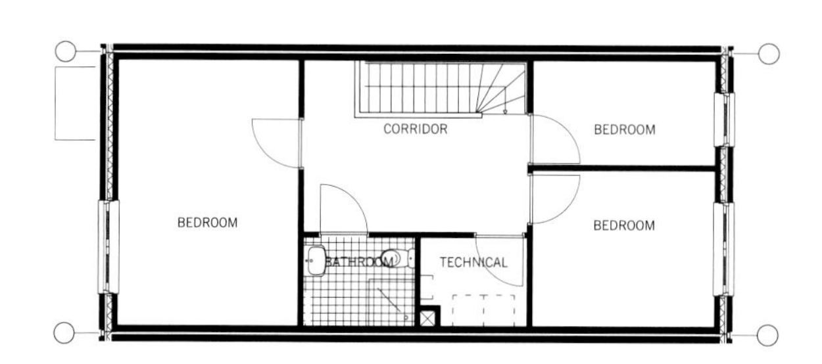

FLOORPLAN LEVEL 1

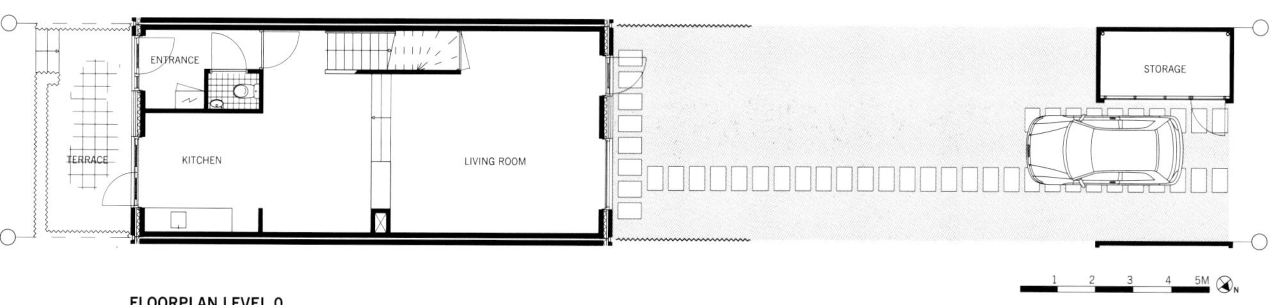

FLOORPLAN LEVEL 0

Profile

Dutch architects of BBVH were looking for a solution to give future buyers of their serial housing projects more influence on the layout and size of their homes. BBVH does not want to design homes only for the average families of 2.31 persons, but we want to design homes that fit. With this basic idea, this project came into being, which includes 30 row houses.

Architectural Design

Having seen one too many floor plans with the same layout, too small to fit a big family, too big for couples without kids the architects thought of a different solution.

Step one was to design a modular house that can easily be altered. Add a room, add a floor, change the layout, enlarge the living etc. All these changes could be made using the same basic module. In this way it became possible to build in the same row a cost efficient house of 135 m^2 and a large five bedroom town house of almost 200 m^2. The architecture of this house is restrained because the diversity of the individual choices will form the final picture. So also the final image of the block is "user generated content".

Step two was to build an online configurator that made it possible for potential buyers to create their own dream house. The software had to be very low threshold and understandable on the one hand but on the other hand it had to contain all relevant and important facts about the house like evaluations of items, square meters, the building plans including the choices made etc. This led to a "step by step" process in which you gradually design your own house.

Room for freedom is the theme of the design for 30 houses in Hoogvliet. With the use of interactive software, potential buyers can compose their ideal home. Freedom of choice is provided in possible extensions for the basic dwelling, both in floorplan and in the number of floors. This results in a diverse facade on an urban level, while the block's identity is ensured through the use of 1 set of materials and finishes. The dwellings all have 2 in stead of 1 main facade with an extra high living room orientated towards a canal.

NEW STONE GATE COMMUNITY WITH FAMOUS SHANGHAI RESIDENTIAL STYLE

19th Century AD Belle Time

海派名居特色 新石库门社区 —— 上海绿地公元1860

项目地点：中国上海市宝山区
开 发 商：绿地集团
建筑/景观设计：水石国际
项目规模：200 731 m²

Location: Baoshan, Shanghai, China
Developer: Greenland Group
Architectural/Landscape Design : W&R Group
Size: 200,731 m²

项目概况

上海绿地公元1860 项目位于宝山区顾村，其住宅定位为高品质联排别墅，总建筑面积为63 078 m²。项目距离地铁1号线共富新村站仅1 000 m，紧邻A20高速出口与南北高架交会处，驱车30分钟内即可抵达市中心。

规划布局

项目在保留传统里弄住宅规划特色基础上，以里坊为单位，将整个规划划分为若干个里坊组团，组团之间为公共活动区域，在延续传统里弄尺度的前提下，选择主弄6 m、支弄4.5 m，保持里弄场所感的同时，满足消防等

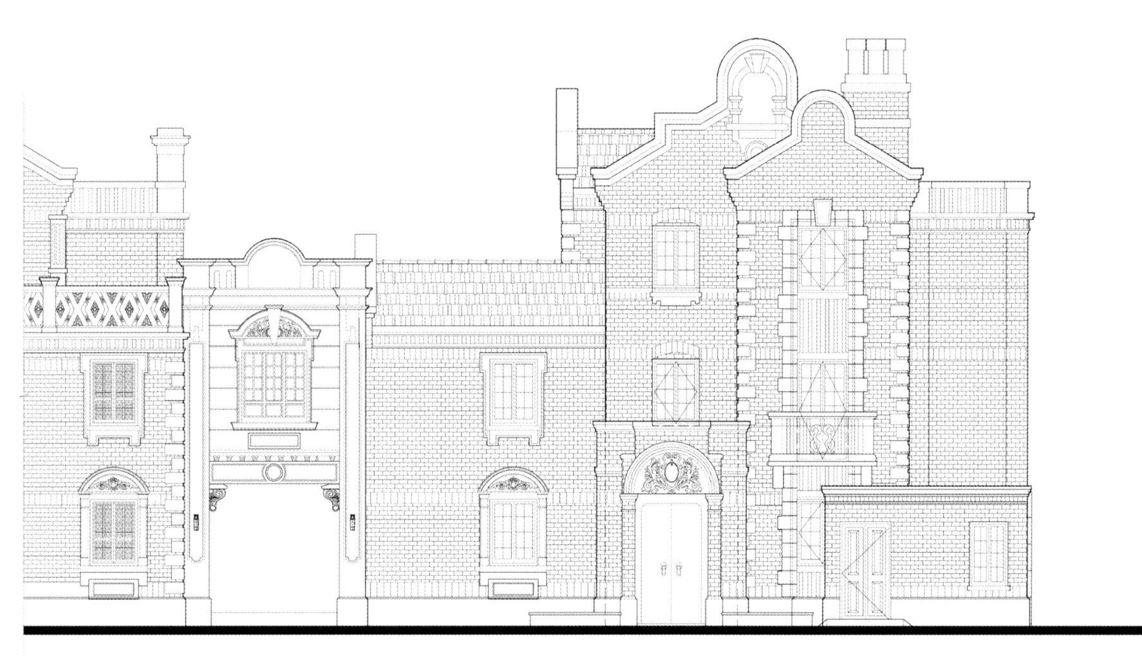

立面图 Elevation

立面图 Elevation

立面图 Elevation

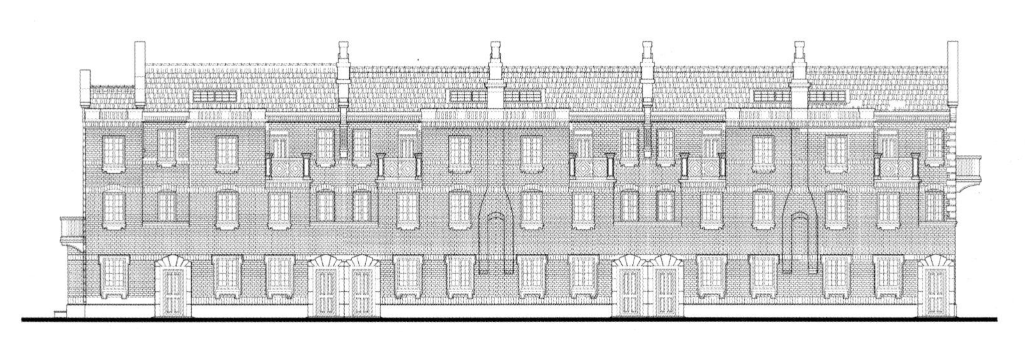

立面图 Elevation

需要。组团结构明晰，空间形态强调归属性，组团的规划布局将每个区域尺度缩小使其更具亲和力，利于交流。同时发挥主弄支弄作为“公共起居室”的作用，充分挖掘里弄中的原有邻里结构、充满活力和亲和力的公共生活。

建筑设计

整个建筑群体高低错落，立面上凹凸变化及窗的不同比例，表现了建筑外观变化和丰富的一面。建筑全面保持里弄空间特色和建筑肌理，在统一的肌理下利用青瓦红瓦、青砖红砖及细部处理等方式，保证里弄形式的多样化，修旧如旧。住宅内院则打破传统封闭方式，把内院与私家庭院连通，精心推敲间隙尺度，创造宜人温馨的小庭院空间。同时通过庭院，使每层主要居室包括起居室和卧室均获得南向采光。

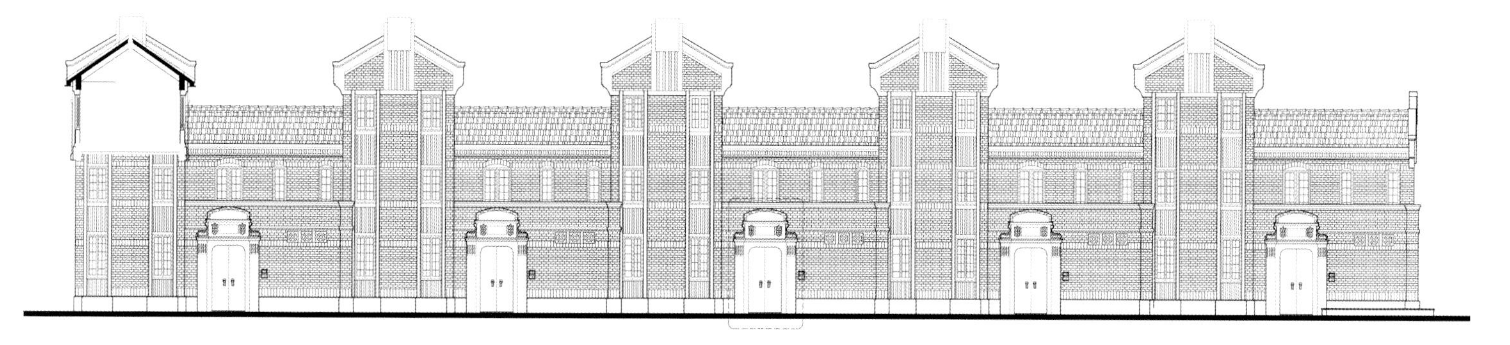

立面图 Elevation

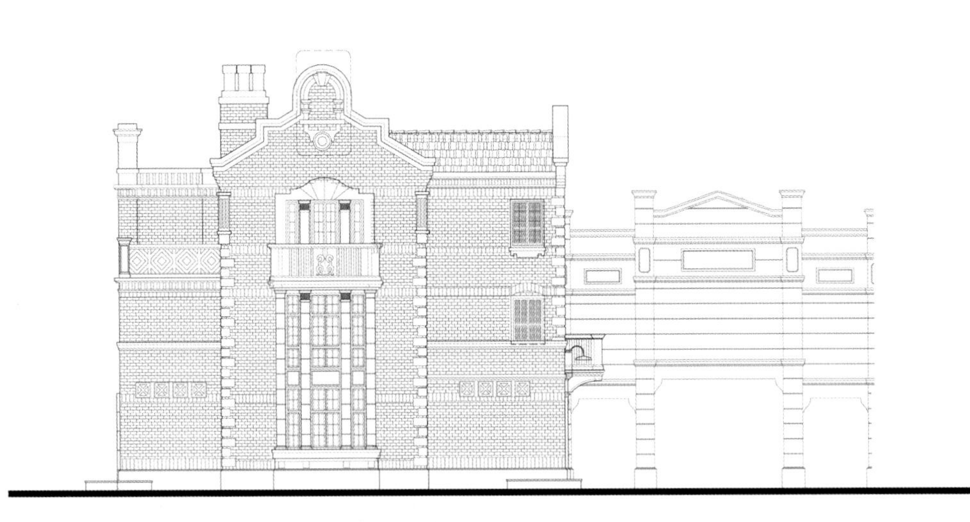

立面图 Elevation

景观设计

项目本身的独特造型就是一道美丽的风景线，红砖墙面加上灰褐色的屋顶再配以灰白色的石库门，让整个项目首先在视觉上给人以美感；在传统的里弄和里坊的社区元素中，种植了具有上海当地特征的景观树木，让里弄与这些绿化景观以及生活在这其中的人们有一种互动的空间，同时能够唤起对传统海派民居建筑的印记。

Profile

Located in Gucun, Baoshan District, the project is set with high-quality townhouses associated with a total construction area of 63,078 m^2. It is only 1,000 m to subway Line 1 Gongfu New Village station, close to A20 highway exit and north-south elevated interchange within 30 minutes' drive to the city center.

Planning Layout

Based on traditional neighborhood residential features, it reserved alleys and divided the whole area into several groups. Under the premise of following the traditional neighborhood scale, public area between the groups select the main lane of 6m and support lane of 4.5m, which kept the neighborhood sense and met the needs of fire protection. Clear group structure, the spatial form emphasis on belonging and the planning layout of the groups narrow the regional scale to make it more user-friendly and conducive to exchange. At the same time, the main lane played the role as "pubic living room", making the best of original neighborhood structure and the public life full of vitality and affinity.

Architectural Design

Scattered groups, bump changes on the facade and windows in different proportions showing the architectural changes in appearance and the rich aspect. Buildings maintain a comprehensive neighborhood space features and architectural texture to ensure the diversification of forms through gray and

red tile and brick and the detail treatment. The inner garden breaks the traditional closed way to connect with private courtyard and create a pleasant and cozy courtyard space through carefully considering the scale. Main rooms including bedrooms and living rooms on each floor have access to light from south.

Landscape Design

The unique shape of the project itself is a beautiful landscape. Red brick walls with taupe roof and off-white stone gate give a sense of beauty at first sight. Landscape trees with Shanghai local characteristics planted in the community with traditional style provide an interactive space between the alleys and the green landscape and the residents there, and arouse the memory of traditional Shanghai Architecture as well.

集资里

兴业里

旅游规划是创造性与实践性的结合体

——访旅游规划设计专家 陈南江博士

■ 人物简介

陈南江
广东省社会科学院旅游研究所总规划师
广东中建设计有限公司董事长
旅游规划博士，研究员，国内知名旅游规划设计专家

1996年从北京大学毕业到广东省旅游局资源开发处工作，2000年组建广东省旅游发展研究中心，曾任主任、总规划师，2009年改任广东省社会科学院旅游研究所总规划师。具有丰富的旅游区规划设计工作经验，擅长从操作层面解决旅游景区度假区规划实务，被誉为南方实战派旅游规划专家的代表，在旅游专业网站上，被誉为中国旅游规划"十大代表人物"之一。

■ Profile

Chen Nanjiang
Chief Planner of Guangdong Provincial Academy of Social Science Tourism Research Institution
President of Guangzhou Zhongjian Design Co., Ltd.
Doctor of Tourism Planning, researcher, domestic well-know tourism planning and design experts

Chen came to Guangdong Provincial Tourism Bureau after graduating from Peking University in 1996, and set up Guangdong Provincial Tourism Development Research Center in 2000. In 2009, he was the chief planner of Tourism Institute of Guangdong Provincial Academy of Social Science. With rich tourist area planning and design work experience, he is good at planning practice from the perspective of operation. In addition, he is known as tourism planning expert on behalf of southern practical school and is honored as one of the top ten tourism planners on tourism professional website.

《新楼盘》：谈谈您眼中的旅游规划是一项怎样的工作？

陈南江：旅游规划的任务是打造旅游产品，发展旅游产业。它是一个创造性和实践性很强的工作。旅游规划设计不是流水线可以生产的产品，每个项目的资源类型和开发条件不一样，文化背景不一样，市场情况也不一样。规划设计团队的负责人非常重要，他必须有丰富的经验，有良好的功底，有很强的责任心。所以，不是规模大、名气大的单位就可以做出好的旅游规划。

规划设计是一个非常辛苦的工作，同时，也是一个很有成就感的工作。用我个人的话来概括，就是："走遍万水千山，几度衫湿衫干。白天鞋底磨穿，夜晚挑灯鏖战。你我规划人员，苦乐自在心间。指点江山开颜，阅尽春色人间！"

《新楼盘》：如今众多的旅游度假区在建成之前都会有相关的规划设计，您觉得在规划的过程中最主要的任务是什么？旅游规划师在这一过程中扮演着怎样的角色？

陈南江：最主要的任务有三个：一是进行客观全面的发展条件分析，二是进行准确的细分定位（市场定位、目标定位、功能定位、规模定位、档次定位、形象定位），三是建立清晰可行的盈利模式。旅游规划师对于这三个方面负有至关重要的责任。

旅游规划师必须对于吃透这个地方（本土文化、自然条件、市场消费习惯、消费能力），吃透该类产品，吃透各类市场，在这三个吃透的基础上进行精心的策划。至于内部的项目布局及基础设施安排，那倒不是重要的，城市规划师可以解决这些后续的问题。建筑设计师和景观设计师的任务是将旅游规划师的思想落实到具体建设形态上。

《新楼盘》：旅游度假区作为旅游主要的目的地，应该怎样去实现社会效益、生态效益以及经济效益的和谐统一？

陈南江：简而言之，旅游度假区应该以生态保护为前提，以市场为导向，发展健康的文化娱乐，兼顾社区居民利益。

《新楼盘》：作为旅游规划方面的知名专家，请您给年轻的规划设计师们一些建议？

陈南江：首先要有两点意识：一是提升专业素质。年轻的规划设计师们以前可能接受过很多年的学校教育，甚至可能有了一些实践经验，但是还需要不断学习。每一个项目都面临不同地域的文脉和市场，网络、杂志和书籍都是学习的手段，有条件的时候要进行现场考察。作为旅游规划设计人员，必须对于国内外的代表性旅游企业和产品有基本的了解，才能够很好地把握规划设计。

二是保持责任心。规划设计的每一笔都将变成客户的投资，都将变成钢筋水泥或者其它物品，如果对于工作不细心研究和谨慎下笔，可能会导致建筑物的坍塌，导致房屋被洪水浸泡，导致人身安全事故，导致游客使用极不方便，甚至导致投资者巨大的亏损。所以，要带着责任心工作，今后到达一个自己参与规划设计的酒店或者度假村去参观的时候，能够自豪地跟同去的朋友或者后辈说："这里是我规划设计的"，而不是听着别人的批评，内心暗暗惭愧，羞于对别人说是自己规划设计的。提交的规划设计文本，更不应该有低级的错误，例如停车场规模、餐厅规模与客房规模远远不配套，甚至在广西的项目里冒出来一个内蒙古的地名等。如果没有良好的专业素质和责任心，就不是一个合格的技术人员。无论在哪里工作，都不可能取得成功。

旅游规划设计人员的成长，首先要达到本专业的"深度"，然后再努力延展知识面的"广度"，一纵一横铺设好之后，再努力结合实际，达到一定的"高度"，才能成为一个专家。每做一个规划，规划设计师都要有两次转身：一是转身为游客，推敲方案是否可向往，可停留，可消费，可回味；二是转身为投资商，推敲方案的可操作性和可盈利性。要努力做到四个满意：群众满意、政府满意、游客满意、企业满意。

《新楼盘》：旅游规划设计对于旅游业来说有何重要的意义？

陈南江：一个行业的发展，核心是产品。对于旅游行业来说，它的主体是三块：观光旅游、度假旅游和商务旅游，对应的主要产品就是景区、度假区与宾馆酒店。旅游规划设计就是打造产品的工作，没有好的规划，就没有好的产品，就难以成功营销，难以取得良好的经济效益，旅游企业和旅游产业都难以发展。

《新楼盘》：您觉得未来旅游规划设计以及旅游业的发展趋势是怎样的？

陈南江：旅游业是永远的朝阳产业，在总体规模稳步上升的背景下，它的内部发展将呈现以下趋势：观光旅游比重下降，度假旅游与体验旅游比重上升；自助旅游比重上升，团队旅游比重下降；大众旅游比重下降，高端旅游迅猛发展；近程旅游比重下降，中程与远程比重上升。

旅游规划设计要因应旅游市场变化，同时又要引领市场消费。未来的旅游规划设计，将越来越做到四个同步：与国际潮流同步，与政策动向同步，与市场趋势同步，与新材料新技术发展同步。

Tourism Planning is the Combination of Creativity and Practicality

——Dr. Chen Nanjiang, Tourism Planning Design Expert

New House: Tell us about tourism planning, what kind of work is it?

Chen: Tourism planning is to create tourism products and develop tourism industry. Tourism planning design is no line products, different types of resources and development conditions of each project is differ in cultural background and market condition. The person in charge of planning and design team is very important, he must have extensive experience, a good foundation and a strong sense of responsibility. Then it doesn't mean every company in large scale and with fame can make a good planning.

Planning design is a very hard work, a hard work with a great sense of accomplishment. In my own words: traveled the long and arduous journey with several shirts wet and shoes worn out; work until deep into the night, these what we planners have to endure, after all pleasant scenery will bound to come to our eyes after the sweet torture.

New House: Now a large number of tourist resorts have relative planning design before completion, what's the most important task in the planning process? What's the role the tourism planners play in the process?

Chen: There are three most important tasks. Firstly, analyzing the comprehensive development conditions objectively; secondly, accurate positioning of the segments (market positioning, targeting, function orientation, positioning of scale, grade positioning, image positioning); thirdly, establishing a clear and feasible profit model. Tourism planners have a crucial responsibility for these three aspects.

Tourism planners must thoroughly understand this place (the local culture, natural conditions, market consumption habits and consumption power), understand these products and the various types of markets, and then plan carefully on the basis of the thorough grasp. As for the internal project layout and infrastructure arrangements, there are not that important and can be solved in the follow-up. Architects and landscape architects' task is to implement the idea of the tourism planners into specific construction form.

New House: As a tourist destination, how should tourism resort do to achieve the harmonious unity of social benefit, ecological benefit and economic benefit?

Chen: In short, ecological protection is the precondition, market-oriented developing healthy culture and entertainment and taking the interest of the community residents into account.

New House: As a well-known expert in tourism planning, could you please give some suggestions to the young designers?

Chen: Two consciousnesses must be pay attention to. Firstly is professional quality. Young designers may have previously received a lot of years of schooling and may even have some practical experience, but they also need to learn. Each project is facing a different geographical context and market than others. Network, magazines and books are means of learning, when conditions permit, to conduct site visits. As a designer, you must have a basic understanding of domestic and foreign tourism enterprises and products, so as to be able to grasp the essence of design.

The second is to maintain a sense of responsibility. Each of the planning and design will become the customer's investment, will become reinforced concrete or other items, with no careful study, it may lead to the collapse of the building and resulting in personal safety incident and even lead to huge losses. So to work with a sense of responsibility, when reaching a hotel or resort with friends that you have been participated to work with, you'd better be the one who is proud to tell you friends that "that's my work", but not the one who ashamed to tell others their own planning and design. Stupid mistakes shouldn't be presented in submitted texts, such as car park scale doesn't match with restaurant size and room size, Inner Mongolia Place names emerge in the project in Guangxi. It is not a qualified technician without a good professional quality and sense of responsibility. No matter where he works, he is unlikely to succeed.

Regarding to the growth of tourism planning designer, professional depth is the first to reach, then extend the breadth of knowledge, combine with practicality. Only by reaching a certain height can you be an expert. Designers should have turned twice in each plan: firstly, turn around for the tourists, deliberating whether the place worth to visit, to stay and to be memorable or not; secondly, turn for investors, considering the operability and profitability. Make efforts to achieve the satisfactions of the masses, the government, the visitors and the enterprises.

New House: New House: How much the tourism planning and design means to tourism industry?

Chen: Products are core to the development of an industry. For the tourism industry, it consists of sightseeing tour, vacation tour and business tour, whose corresponding products are scenic spots, resorts and hotels. Tourism planning and design is to create products, no good planning no good products, no successful marketing no good economic benefits, tourism enterprises and industry will be difficult to develop under such circumstance.

New House: How do you think of the future of tourism planning design and tourist industry?

Chen: Tourism is always a sunrise industry. In the context of a steady increase in the overall size, its internal development will show the following trends: share of sightseeing tour declines, proportion of vacation tour and experience tour rises; the proportion of alternative tourism rises, group tour falls; mass tourism declines, high-end travel boosts; short-range tourism declines, medium and long-distance tourism rise.

Tourism planning and design change along with the tourism market and lead the market consumption at the same time. In the future, it will reach four synchronizations little by little: synchronized with the international trend, policies, market trend and the development of new materials and technology.

DELICATE PHILOSOPHY OF LIFE, SIMPLE COURTYARD

| Country Garden Holiday Islands Flowers Huadu Third Street

精致的生活观念 简洁的庭院语言

—— 花都碧桂园假日半岛鸟语花香三街

项目地点：中国广东省广州市花都区
景观设计：广州市圆美环境艺术设计有限公司
总占地面积：6 670 000 m²

Location: Huadu District, Guangzhou, Guangdong, China
Landscape Design: Guangzhou Yuanmei Design
Total Land Area: 6,670,000 m²

此设计是现代式庭院，主要讲究的是一种简约之美，追求自由、奔放和大气，大则幽深静谧，小则精致秀巧。

功能设计等方面营造出一个舒适、绿色、健康、富有文化气息和地方艺术特色的居住环境。

住宅前的庭院，是家庭向外界过渡的地带，也是家庭空间的延伸。小小庭院好处多，一个环境好功能齐全的庭院，不仅可以让人们足不出户就可以轻松回归自然，也可以与家人或朋友坐在其中用餐、聊天，放松心情。

景观元素主要是采用简单的长方形、圆形和梯形等，既美观大方，又不乏实用性。庭院中的构筑物形式简约，采用抽象雕塑品、艺术花钵等为庭院的主要装饰元素，同时庭院中还将运用一些天然的元素，例如石块、鹅卵石、木板等。庭院绿化采用垂直高大的乔木形成高大、狭长的线条，配置低矮的植物来达到视觉上的平衡。此外还使用竹子等植物装饰局部景观。

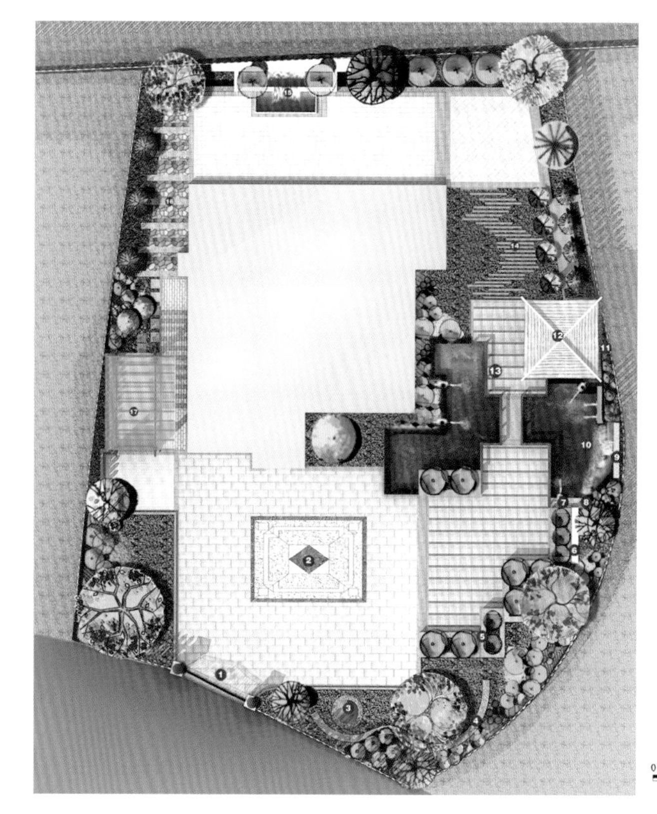

1 别墅入口
2 拼花铺装
3 吐水小景
4 弧形矮景墙
5 错级花池
6 文化石景墙
7 吐水小品
8 散置鹅卵石
9 跌水景墙
10 深水鱼池
11 亭背景景墙
12 现状凉亭
13 眺水平台
14 条形汀步
15 多级跌水景
16 过渡空间
17 上抬花架

景点总平面图 Scenic Spots Plan

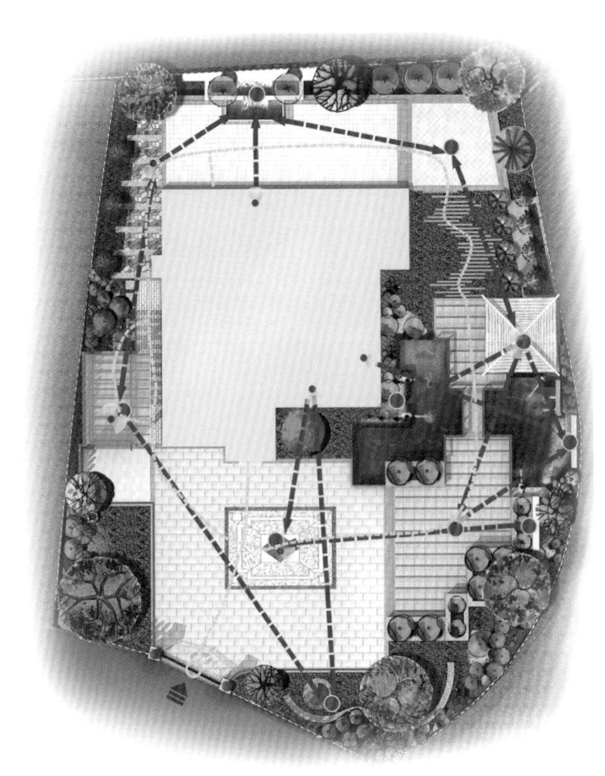

景观视线分析图
Landscape View Analysis

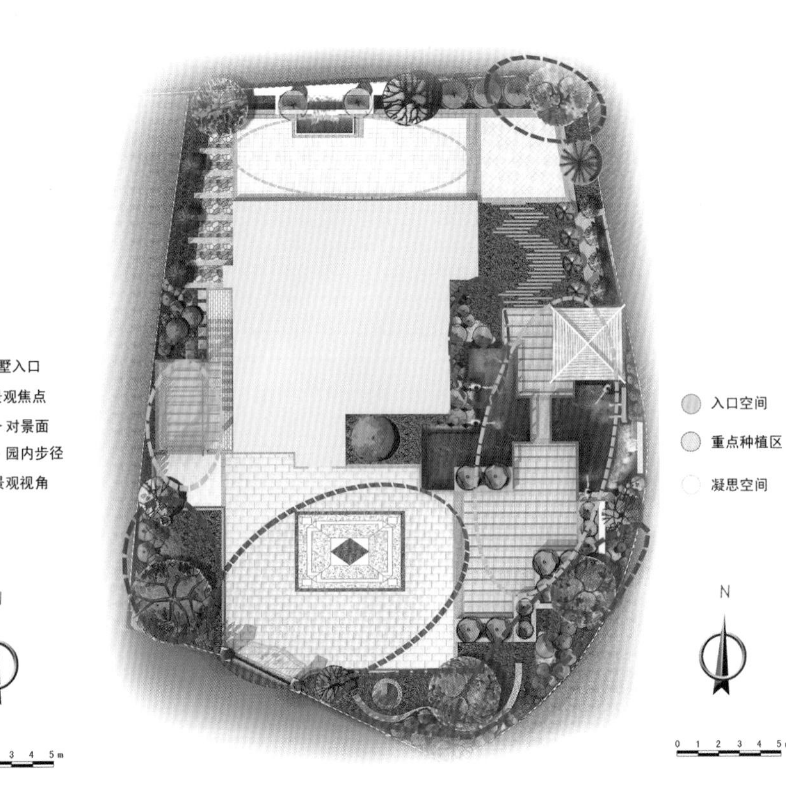

功能分析图
Functional Analysis Drawing

This is a modern-style courtyard which emphasizes on a simple beauty and pursues for free, unrestrained and generous atmosphere.

In terms of functional design, it creates a comfortable, green and healthy living environment featuring both rich cultural atmosphere and local artistic characteristics.

Courtyard in front of the residence is not only the transition zone to the outside world, but also an extension of the family space. A full-featured garden with friendly environment not only allows you to feel the natural world without stepping out, but also dine, chat with family or friends inside.

Regarding to landscape elements, simple rectangular, circular and trapezoidal looked

beautiful with no lack of practicality. Abstract sculpture, art flowerpot are the main decorative elements in the garden, besides, there are natural elements such as stones, pebbles, wood, etc. Tall and narrow lines formed by tall trees configure with low plants to achieve visual balance. In addition, bamboo and other plants are used to for decoration.

INHERITING REGIONAL CONTEXT, EMBRACING MEIHU SCENERY

Vanke Rain Garden, Nanchang

传承空间文脉 揽尽梅湖风景—— 南昌万科润园

项目地点：中国江西省南昌市
开 发 商：江西省万科益达房地产发展有限公司
景观设计：深圳市雅蓝图景观工程设计有限公司
占地面积：130 000 m²

Location: Nanchang, Jiangxi, China
Developer: Jiangxi Vanke Yida Real Estate Development Co., Ltd.
Landscape Design: Shenzhen ALT Architectural Landscape Design Co., Ltd.
Land Area: 130,000 m²

本项目位于江西南昌青云谱传统商业街区一侧，原为公安学校用地，用地内有若干多年生长的大树，林荫婆娑，枝桠虬然，宛若水墨意境。本项目的景观便从此落笔，配合建筑规划，因借风景资源，在风景之中生长风景，在文化之畔坐落生活。

景观设计方面，项目传承空间文脉特征，通过层进的庭院空间围合连接不同的功能与景观空间，在递进的层次中转换从城市公共空间到私家专属空间的气质氛围与心理感受，由此营造出具有品质感受的家园景观。尊重场地内的现有风景资源，保留原生大树，并作为社区景观的主要元素。合理布局社区功能空间，既满足户外生活的功能要求，又保证居住的安静与私密。有效处理场地竖向高差，并由此产生新颖的景观效果。兼顾东、西区景观的呼应与联系，使整合的风景联络被分隔的社区。

项目在植物景观配置上，充分结合狭长的南北走向场地特点，尊重场地内的现有风景资源，保留原生大树，希望真实地塑造出绿意葱茏的绿色居住环境。以适地适树为原则选用本土特色植物，结合现有大树及观赏大灌木、丛生地被灌木与花卉进行立体层次营造，自然中蕴含野趣，清新处不失质朴，现代而延续文脉。林木种植以自然的树丛结合为主，在统一的节奏中变化出自然的种植形态，在狭窄有限的空间中营造出丰厚的植被感受及幽深的环境体会。设计中同时重视植物景观四季季相的变化特征，景随季变的同时形成丰富的植物空间群落，给人以回归自然的清新感受。

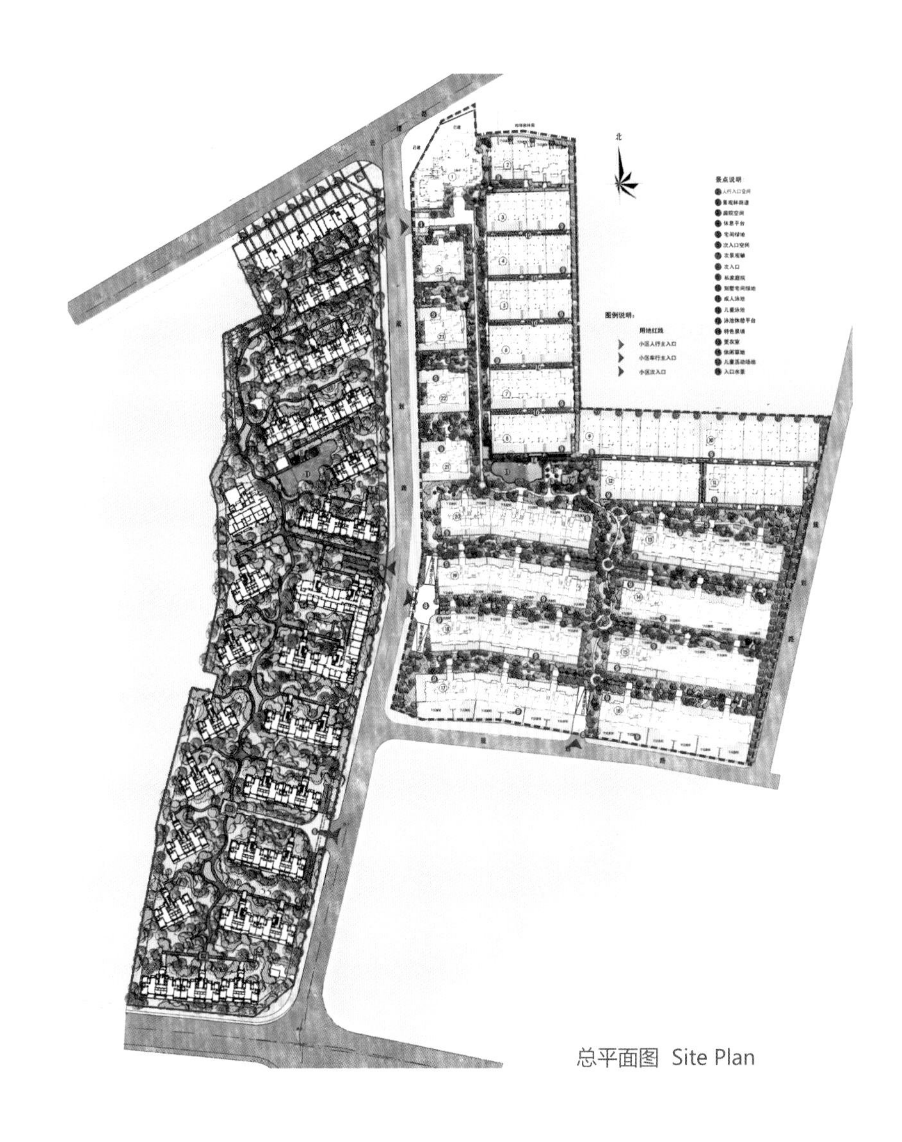

总平面图 Site Plan

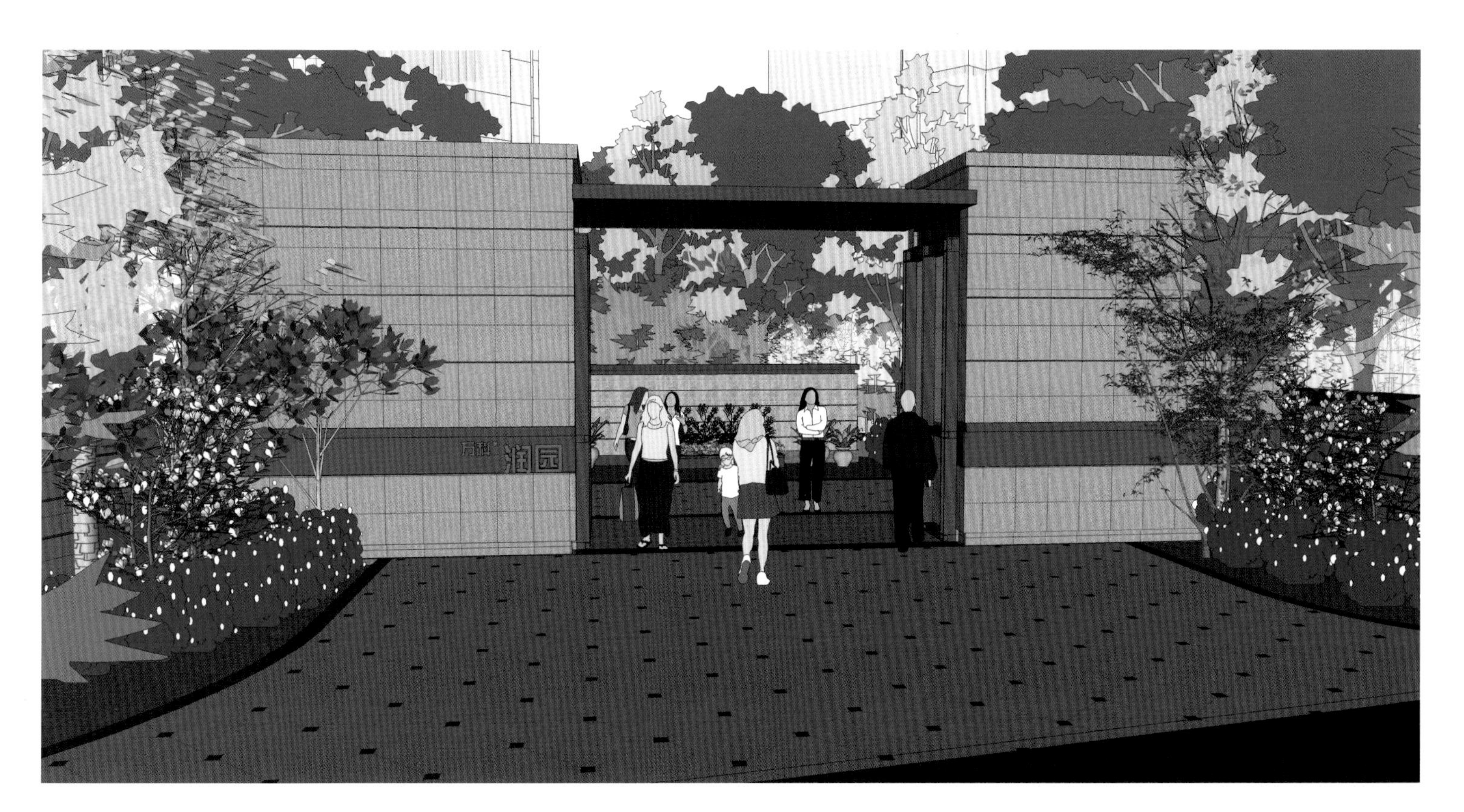

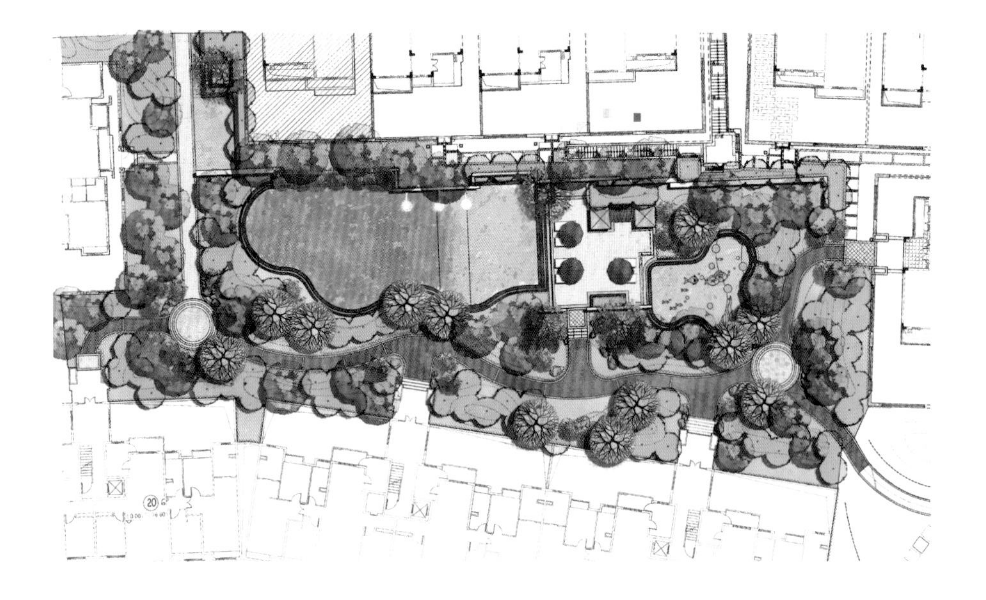

Located in Qingyunpu traditional commercial district, former of the site is for public security school. Trees exist for several years set a great start for the landscape design of this project which cooperates with architectural planning and makes the best of the scenery resource.

In terms of landscape design, it inherits regional context characteristics, connecting different functional and landscape spaces through layering courtyard space, converting atmosphere and mental feelings from urban public space to private space in progressive levels, thus to create fine quality home landscape. The proper functional layout not only meets the need of outdoor activities but ensures the privacy. Vertical height difference has been processed effectively

小区入口立面图 Entrance Elevation

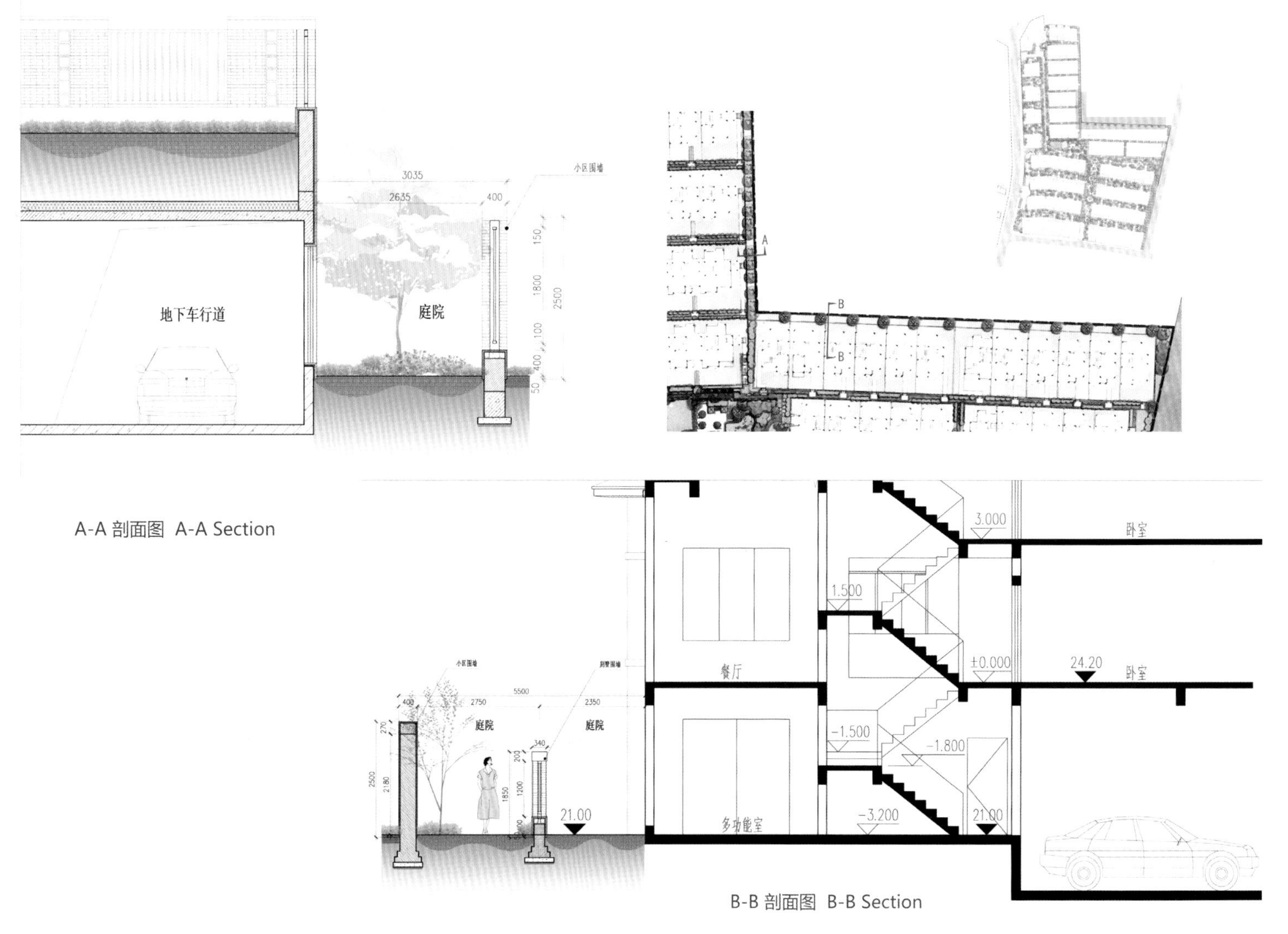

A-A 剖面图 A-A Section

B-B 剖面图 B-B Section

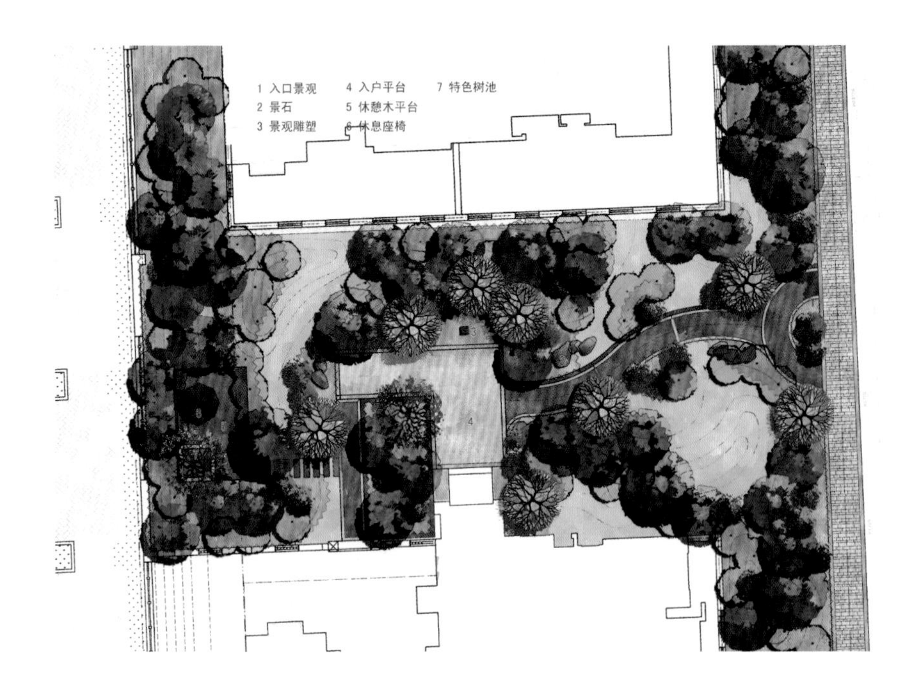

that produces a novel landscape effect. It considers the relation of the landscapes in both east and south areas, connecting integrated scenery with divided community.

In terms of plant landscape configuration, it fully integrates with the characteristics of the north-south narrow site, respecting existing scenery landscape and retaining original trees to green living environment in a true sense. It selects local green plants in according to the conditions of the site and the trees, creating three-dimensional levels by integrating trees, brushes and flowers to express the natural flavor and inherit regional contexts. Natural planting form and profound environment experience came into being in the unified rhythm and limited space. In addition, it pays attention to the seasonal variable characteristics of the plants, in which you can see the fresh green plants all the year around.

香樟林大道

NATURAL AND ECOLOGICAL SOUTHEAST ASIAN LANDSCAPE

| The Concept Planning of Yangjiang Spring Resort

自然生态的东南亚风情景观—— 阳江温泉度假村概念规划

项目地点：中国广东省阳江市
客　　户：阳江市南湖国旅凤凰湖国际温泉度假村开发有限公司
景观设计：奥雅设计集团
占地面积：4 000 000 m²

Location: Yangjiang, Guangdong, China
Client: Yangjiang Nanhu Travel Phoenix Lake International Hot Spring Resort Development Co., Ltd.
Landscape Design: L& A Design Group
Land Area: 4,000,000 m²

阳江位于广东省西南沿海，有着丰富的旅游资源，依山傍海，自然旅游资源品种全、品位高、空间组合佳。该项目拥有4 066 666.67 m^2可用地，为创造规模性景区和特色景区提供了机会。本案设计的目标是打造一个东南亚风情的国际五星级温泉度假村，利用当地的天然温泉资源，打造一系列以水为主题的国际5星级度假休闲天堂，带动当地的旅游经济发展，展示阳江的自然生态与当地文化，为建立良好生态环境的新城提供了机会。

通过与水相关的主题来体现亚洲水文化，成为阳江的一个亮点名片，对此处景观的开发可展示阳江本土文化，带动当地旅游业发展的星级旅游热点，另外，也借此建设一个接待国际宾客、商人、行政人士等的高级优质服务中心。

本案以生态为主题、运用东南亚风情的设计语言，综合度假村开发。凤凰涅槃，浴火重生，暗喻基地的开发将如山鸡蜕变成凤凰。无论是整体的形态或是使用功能，都是一种质的改变。其建筑形态独特，象征凤凰飞舞，与度假村主题风格相一致。

本项目植被茂密，具有景观性，最突出的一点是充分利用天然的水资源来实现亚洲水文化中所蕴含的健康、环保理念。如饮茶、有机食品、spa、温泉等。利用湖区和小水体所提供的丰富的水资源，把水作为总体规划的主导元素，形成湖、溪流、池塘、湿地、温泉等不同水景。在同一景区中创造不同的景点作为其独特的卖点，体现公共空间中的个性。

本案遵循低碳交通运作原则，在旅游区域内使用电瓶车、自行车，充分强调环保低碳的理念。

Located at southwestern coast of Guangdong Province, Yangjiang is rich in tourism resources with wide variety, high quality and great spatial organization, and near the mountain and by the river. This project covers an area of 4,066,666.67 m^2 which provides a chance to create large scale and special scenic area. It aims to make an international five-star hot spring resort in Southeastern Asian style by exploring local hot spring resources, which develops local tourist economy, shows local natural ecology and culture and establishes a good ecological environment for the new city.

It is one of Yangjiang's inviting cards to reflect Asian water culture through water-related themes. In addition to interpret local culture and drive the development of local tourism, this design build a high quality service center for welcoming international guests, businessmen and officers.

Themed as ecology and decorated with Southeastern Asian style, qualitative changes occurred in terms of overall form and function. The unique phoenix-like architectural form corresponds to the theme and style of the resort.

The most prominent in the luxuriant vegetation area is that natural water resource is in full use to realize the healthy and environmental concept that the water contains in Asian culture, e.g., drinking tea, organic food, spa and hot spring. Water is the dominant element in the overall plan with the help of rich water resources, thus waterscapes like lake, stream, wetland and hot spring produced under such a circumstances. Different scenic spots stand a unique space in different scenic areas.

In accordance with the low carbon traffic operation principle, battery truck and bicycles are used in the tourist area to emphasize the idea of environmental protection.

旅游度假区

专题导语

在我国，旅游度假区正处于从观光旅游为主向观光、度假、专项旅游并重发展的转折期，旅游度假区为适应旅游项目而呈现出多种多样的建筑风格和模式，如出现了城市型、人文历史型、生态自然型、休闲娱乐型等形态的旅游度假区。禅都文化博览园修建项目、秦皇岛植物园项目、千岛湖润和度假酒店项目、唐山湾祥云岛项目、狮子湖阿拉伯会议酒店项目等分别展示了旅游度假区多种形态的特色和特点，可细致鉴赏。

Introduction

China is now during the transformation period from pure sightseeing tourism to the tourism integrating sightseeing, vocation tour and special tour. The tourist resorts thus have presented a great variety of architectural styles and patterns to adapt to new tourism programs. For example, there appears tourist resorts of urban style, cultural and historical style, ecological and natural style, leisure and entertainment style, etc. Here we have selected projects of different types and styles, showcasing different features and characteristics of the tourist resorts. The cases, including "Constructional Detailed Plan for Zen Exposition Park, Yichun", "Qinhuangdao Botanic Garden", "Thousand Island Lake Runhe Resort Hotel", "Tangshan Xiangyun Island", "Lion Lake Arab Conference Hotel, Qingyuan", will be great for appreciation.

北京国泰上城置业
BEIJING GUOTAI SHANGCHENG PROPERTY

企业概况

北京国泰上城置业有限公司坐落于北京国贸CBD中环世贸大厦，公司经营范围涵盖旅游地产、住宅、商业、文化体育产业、钢铁、建筑及装饰材料的经营等领域。目前公司下辖廊坊市上城房地产开发有限公司、三河山海树房地产开发有限公司、唐山新戴河旅游开发有限公司等三个项目公司，形成“山海树”的战略布局。公司下属四个控股子公司：山海树股份有限公司、北京山海树规划建筑设计有限公司、北京山海树高尔夫运动管理有限公司和北京山海树林泉书画院，建立并逐渐完善了包含地产、高尔夫管理、物业管理、酒店管理、规划建设、书画院、钢铁等行业的经营格局。

产品与服务

公司致力于打造中国高品质、高品位的高端旅游地产，成为高端旅游度假产品的全方位供应商。公司坚持 “产品质量是企业的生命”，产品从设计一开始就挑选全球顶尖的设计公司，公司在全球范围内挑选著名的合作伙伴，先后与新加坡悦榕集团、伊顿集团、美国易道设计公司等签署合作协议或战略合作协议。

Profile

Beijing Guotai is located in Beijing Guomai CBD Central World Trade Center, whose business scope covers the field of tourism real estate, residential, commercial, cultural and sports industries, iron and steel, construction and decoration materials, etc. At present, it governs three companies, Landfang Shangcheng Real Estate Development Co., Ltd., Sanhe Shanhaishu Real Estate Development Co., Ltd. and Tangshan Xindai River Tourist Development Co., Ltd., and forms the "Shan Hai Shu" strategic layout. It subordinates four wholly-owned subsidiaries, Shan Hai Shu Company Limited, Beijing Shan Hai Shu Planning & Architectural Design Co., Ltd., Beijing Shan Hai Shu Golf Sports Management Co., Ltd. and Beijing Shan Hai Shu Linquan Calligraphy and Painting Academy established and gradually perfected the operating pattern including real estate, golf management, property management, hotel management, planning and construction, Calligraphy and Painting Academy, iron and steel, etc.

Product and Service

The company committed to creating high-quality, high-grade and high-end tourism real estate, becoming an integrated solutions provider for high-end tourist resort. Adhering to "product quality is the life of enterprise", it selects the top architecture firms from the beginning of design and then selects well-known partners. The company had strategic cooperation with Banyan Tree, Eaton Group and EDAW.

宜春城市投资集团

YICHUN CHENGTOU GROUP

企业概况

宜春市城市建设投资开发总公司是国有大型企业，二级资质。现有资产84亿元。公司设五部三室八个职能部门，下辖房地产、国有资产经营管理、拆迁、物业等四个子公司，七个工程项目派出机构。公司成立于2001年，公司共融资近50亿元。

产品与服务

公司先后建设完成了朝阳路、宜春大道、中山西路改造、袁河西路、清沥江大桥、迎宾公园、春台公园西大门、秀江东路一期、宜阳路改造、青山路、机场B线、袁山大桥等20多项大、中型工程。打造了一批城市亮点，为宜春举办全国第五届农运会、创建全国园林城市、全国卫生城市、中国优秀旅游城市，为宜春的城镇化建设及经济发展做出了贡献。2010年，公司全力以赴投身城市建设·“三年大会战”，共投资12亿多元，正在紧锣密鼓建设宜阳大厦、四馆一社、环城西路改造，卢洲大桥及卢洲南路、秀江东路二、三期以及明月大道、袁河西路、北湖、欧式风情一条街等建设开发项目。

Profile

Yichun City Construction Investment & Development Corporation is a large state-owned secondary qualification enterprise with existing assets of 8.4 billion. It consists of five units, three sections and eight functional departments; four subsidiaries cover real estate, state-owned assets management, demolition and property; and seven engineering project agencies. Founded in 2001, the company has been financing 5 billion.

Product and Service

The company has completed the transformation of more than 20 large-sized and medium-sized projects such as Chaoyang Road, Yichun Road, Zhongshan West Road, Yuanhe West Road, Qinglijiang Bridge, Yinbing Park, Chuntai Garden West Gate, first quarter of Xiujiang East Road, Yiyang Road, Qingshan Road, Airport Line B and Yuanshan Bridge, etc. A number of bright spots it created make great contribution to the urbanization and economic development of Yichun: the fifth National Peasant Games held in Yinchun, and it also devoted to become the national garden City, national hygienic city and China's outstanding tourist city. In 2010, the company made every effort to join the "three-year battle" to build Yiyang Building and a community of four pavilions, reconstruct Ring Road and develop Luzhou Bridge, Luzhou South Road, the second and third quarters of Xiujiang East Road, Mingyue Road, Yuanhe West Road, North Lake and the European style street.

旅游度假的新时代

■ 作者简介

唐艳红
ECOLAND易兰 副总裁
美国风景园林协会会员
美国城市土地规划研究院会员
清华大学EMBA客座教授
北京园林协会常务理事

■ Profile

Jenny Tang
Vice President of ECOLAND
Member of American Society of Landscape Architects
Member of U.S. Urban Land Institute
EMBA Visiting Professor of Tsinghua University
Executive Board Member of Beijing Landscape Architecture Association

二十年前，中国国家旅游局开始正式批准规划建设国家级旅游度假区，当时的设计目的不仅在于促进我国由旅游观光型向观光度假型转变，同时也为了扩大对外开放，吸引外国游客到中国来度假，包括第一批苏州太湖国家旅游度假区、北海银滩国家旅游度假区等共计12项目，虽然在之后的几年间这些度假区大多还没有达到预期的赢利效果，但可以说，这批国家级旅游度假区的建设是中国在度假产品开发上的首次大规模尝试。

如今的旅游度假开发像雨后春笋一般在全国范围内成长起来，规模宏大，项目层出不穷。与此同时，城市的快速发展使得城市中供人们游憩的自然环境越来越难找，人们希望换个新鲜的环境、改变生活节奏、满足个人兴趣以及享受独特的自然风光，因而结合了休闲、度假、娱乐功能的旅游度假区成为人们身心再生的优先选择，生活方式的转变使得人们休闲时间增加，也促使了度假区的大量发展。度假区的题材涵盖了滨海、河湖、山地、滑雪、温泉、乡村、高尔夫、游艇等极为丰富，早已超脱了“上车睡觉、下车拍照”的简单观光模式，目前基本上完成了观光旅游向度假旅游的转变。旺盛的市场需求也使得旅游度假区衍生出旅游休闲及会议会展等多种功能相结合的形式，特别是在国家对城市住宅宏观调控的大背景下，旅游地产正成为众多房地产开发商转型的新热点。

但是在蓬勃发展的背后，旅游度假区的开发也显现出了若干“不和谐”的问题，集中表现在两方面，一是度假开发与自然环境不和谐，项目建设行为强势，更多的是在强行改变场地的自然空间形态，而不是将项目归融于自然；二是度假开发与原住民社区不和谐，度假区的规划普遍缺乏对周边社会经济衔接的考虑，经常导致度假区内外经济发展水平与阶段迥异，宛若两个世界，由此带来了一些社会矛盾，影响了度假区的正常运营。越来越多的事实证明，这样的度假开发模式是无法持续的，尤其是在人多地寡、人均资源稀缺且处在发展中阶段的中国。

值得欣慰的是，业内外众多人士开始注意到了这些迹象，并做了积极的研究与思考，虽然国内设计界对旅游度假区规划设计的研究在整体上还处于初期阶段，一些的设计机构已经在成功的案例实践中开始理论总结，例如，ECOLAND易兰规划设计团队在海南海口江东红树林国家生态湿地度假区和保亭县什进村乡村旅游度假区的项目上，针对上述旅游度假区开发的“两个不和谐”问题，在规划设计工作中做了探索性突破。

在红树林国家生态湿地度假区项目中，其场地最大的资源特色为红树林，而在过去的数十年间，由于生产生活的需要，大片的红树林被改造成为鱼塘或其他用途，红树林损毁面积逾半。在易兰的规划中，明确提出了不仅要恢复损毁的红树林群落，还要在条件适宜的场地接续湿地生态景观，并与现存的红树林景观融为一体，为游客提供独具特色的红树林湿地度假景观。与此同时，当地居民赖以生活的产业结构也随着度假区的规划也进行了调整，即由传统小农业升级到高附加值的度假服务业。在这里，度假区的开发为游客、当地居民和大自然提供了一个和谐相处、各得其所的良性发展空间。

在海南呀诺达热带雨林旅游度假区的规划中，区域内保亭县的什进村原是一个普通的黎族村，村舍建筑低矮简陋，全村人均收入在两千元左右。针对这一区域，规划团队提出了“大区小镇”的全新设计理念，将农民、农业、农村作为核心的度假要素纳入统一规划设计，让农民不搬迁而变成黎族乡村民俗文化的使者，农业变成怡人的稻田度假景观，而村舍则改造成为具有浓郁地方文化特色的度假接待设施（船形屋稻田酒店），游客、政府、开发商、原住民真正实现了四方共赢，旅游度假区的发展与当地社区水乳交融，社区的繁衍生息（起居耕作）为度假区的持续经营提供了源源不断的生机和活力。目前该项目一期已经建成，全村人均收入已突破万元，受到了社会各界的广泛关注。

有人云，二十一世纪是中国人的世纪，其实，这也是中国开始迈向“全民度假”的世纪。有理由相信，只有彻底完全的融入到场地生态环境及区域内世代生息的居民中去，才可能获得无限的创作灵感，做出真正经典的度假作品，正所谓“文章本天成，妙手偶得之”。

New Era for Tourism Planning and Design

Twenty years ago, the National Tourism Administration of the People's Republic of China officially approved the planning and construction of state-level tourist resorts. The objective was not only to promote the transformation from sightseeing tourism to holiday tourism but also to entice more foreign tourists to vacation in China. Though many of the projects - including Suzhou Taihua Lake National Tourism Resort and Beihai Silver Beach National Tourism Resort - did not obtain the level of attraction they initially sought, these projects marked the beginning, the first foray for China to majorly develop its tourism industry.

Recently, tourism projects have been rapidly springing up across the nation, with various scales and forms. At the same time, natural green spaces for recreation and leisure are becoming scarce with the rapid development of cities. People want to have a new environment to change the pace of life, satisfy personal interests and enjoy unique scenery. Thus tourism resorts, integrating the functions of recreation and entertainment has become the first choice for people. This change of lifestyle increases leisure time and promotes the development of the resort areas. As such, the most successful of tourism resorts tend to strike a balance between providing a relaxing escape from bustling everyday life and ensuring various forms of entertainment and activities for travelers to engage in. These resorts are typically centered on an overarching theme, for example: seaside, lakeside, forests and mountains, skiing, hot spring, village, golf, and so on. Modern tourism has evolved from simplistic sightseeing and the old routine of 'sleeping on the tour bus and arriving to take photos.' The strong surge in market demand for tour, leisure, and convention or exhibition, in combination with the adjustments of China's housing market have made tourism projects attractive ventures for many real estate developers.

With this rapidly burgeoning industry come its subsequent setbacks and behind the face of prosperous development, there are several disharmonies: firstly, the friction between tourism development and the natural environment. The construction is always, to some extent, a kind of invasion on the existing natural environment. Secondly, is the discord between the development area and surrounding areas. The planning for these tourism areas usually lacks consideration to its surroundings on the social and economic scale, forming a large disparity between the tourism area and its surroundings. Sharp contrast forms, bringing social conflictions, affecting the future operation of tourism areas. As China bears a large population with relatively little land and limited resources, these two disharmonies, if unaddressed, can cause lasting ramifications, both social and environmental.

The bright side is that, people have noticed these warning signs and started analyzing and solving this problem. Although the research on the planning and design of the tourism projects has been limited, a few design institutions have summarized some theories from their successful practices. For example, ECOLAND Planning and Design firm has provided new solution to problems mentioned above during the planning for Haikou Jiangdong Mangrove National Eco Wetland Resort and Baoting County Shijin Holiday Village.

For the site of the Mangrove National Eco Wetland Resort, the mangrove forests were the special preexisting natural resources. In the years before, many mangrove forests were transformed for fishponds or repaved for infrastructure and other uses due to the demands for production and living. About half of the mangrove remained. Integrated in ECOLAND's planning, is a rehabilitation program that will not only continue to preserve the ecological wetland landscape, but also work to recover the damaged mangrove forests. The plan includes an integration with the existing mangrove landscape, thereby providing the tourists with unique landscape experience. The local industrial structure is also altered accordingly from the traditional individual farming style to a more profitable tourism service industry. The development of the resort will have thus incentivized a preservation of the existing natural landscape and brought about a symbiotic relationship between tourists, local people and the environment.

Similarly, the Hainan Yanoda Rain-Forest Cultural Tourism Zone, located in Baoting County, Hainan Province, also had a special preexisting landscape: the site was initially a village of the Li nationality. Dwellings, though low and crude were also culturally charming, average income of the villagers was RMB 2,000 per month. Under these situations, the team has put forward a new idea, "big zone and small town", taking agriculture, farmers, and scenic countryside into account as the important elements for large area tourism planning. Instead following the conventional develop model of relocating the locals, razing over the preexisting agricultural society, the team wanted to maintain the old dwellings as a point of interest and incorporate the local community in this tourism endeavor. Some of the farmlands would be reformed into scenic landscapes to increase tourist appeal while the business of other farms would increase in response to growth of the resort. Old dwellings were upgraded into tourism facilities build with Li culture characters. Most importantly, the local people would not be displaced. They stay and become the messengers of Li's folk culture. Therefore, it has been a mutually benefitting for tourists, government, developer and local people. The tourism resort development together with the local communities ensures sustainability of the development. With Phase One of the development completed, the local communities have already seen a rise in their average income from approximately RMB 2,000 to RMB 10,000 per month.

It is said that the 21st century will be the century for China. It will also be the century for "tourism" in China. Thus, we strongly believe the importance to integrate eco tourism into people's life's and creating projects that are sustainable and environmentally coconscious.

Web2.0时代的旅游度假区规划设计新趋势

【摘要】随着信息技术的发展，世界进入Web2.0时代，以微博为代表的基于“用户创造内容—平台分享信息”的架构对旅游活动产生了潜移默化影响，Web2.0时代的旅游呈现需求个性化、时间碎片化、出行常态化、渠道草根化、营销口碑化等新特征，与此相适应，旅游度假区规划设计应转变思路，以应对Web2.0时代的新形势。这些思路转变体现为“蓝海、原创、极化、混合、全时、多赢”等六个新趋势。

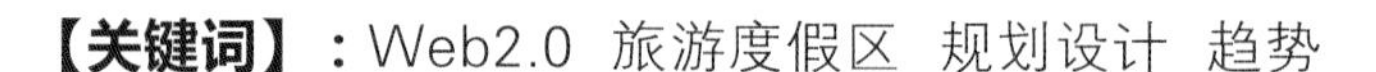

【关键词】：Web2.0 旅游度假区 规划设计 趋势

■ 作者简介

王彬汕
北京清华城市规划设计研究院旅游与风景区规划所所长
博士，高级工程师，国家注册城市规划师

1993年考入清华大学，先后攻读建筑学、城市规划和景观学专业，并获博士学位。毕业后就职于北京清华城市规划设计研究院，现任院务委员会委员、旅游与风景区规划所所长、风景园林研究中心副主任。对风景区、度假区、旅游城镇有深入研究和丰富经验，参与了泰山、黄山、西藏、北戴河、海南国际旅游岛等几十项重大规划项目，并为万达、万科、富力、华润等地产公司提供旅游开发项目咨询。

1、概述

信息技术是当代最伟大的革命之一，当代社会已跨入Web2.0时代，以微博为代表的基于“用户创造内容—平台分享信息”的模式在旅游者之间、旅游者与旅游目的地之间搭建了沟通与互动的桥梁，促使旅游活动产生潜移默化的改变，对旅游度假区的规划建设产生了深刻影响，开发者亟需转变思路，跟上Web2.0时代的旅游新形势。

Web2.0概念2004年由Dale Dougherty和Craig Cline共同提出，并很快成为影响整个互联网界的重要思潮。Web2.0思潮以六度空间、长尾、社会资本、去中心化等理论为支撑，相对于Web1.0，Web2.0的重要特征包括：“用户自行产生内容、共同创作、网络权力的去中心化、利用集体智慧的开放参与体系的构建、长尾市场的崛起、互联网服务的永远beta 版状态等[1]”。“基于Web2.0 的全新经营方式的登场将使得整个旅游业重新洗牌 [2]”。

2、Web2.0时代的旅游新特征

2.1 需求个性化

Web2.0技术的互动特征，实现了供应商与消费者的双向交流，使旅游度假产品的开发更加契合消费者需求，通过在线预订和交流互动提供个性化的定制服务成为可能。旅游者的个性化需求越来越强烈，这导致旅游度假市场的两级分化：一类是传统的大型旅游度假地，提供标准化、大众化的度假产品，依托大型旅游电子商务网站，走规模经济路线；另一类是越来越多的新兴小型旅游度假地，提供个性化、精品化的度假产品，专注于细分市场，走体验经济路线。对于Web2.0时代兴起的第二类市场来说，黄山德懋堂（@德懋堂，@德懋堂卢强）和北京皇家驿栈（@皇家驿栈酒店，@皇家驿栈刘少军）都是值得借鉴的成功案例，两家精品度假酒店借助特殊地理位置、地域特色设计和个性化体验，在高端市场占据了一席之地。

2.2 时间碎片化

Web2.0时代是信息“碎片化”的时代，传统专注式的单线程思维模式，逐渐被碎片化的多线程思维模式替代，旅游者行为也同样呈现某些“碎片化”特征。Web2.0时代的旅游者注意力更加分散，容易被某一旅游意象，某一热点话题所吸引，进而触发旅游度假冲动。“背起行囊就出发”、“周末到巴黎喂鸽子”等意象成为一种惬意的生活符号，旅游度假者的出行时间不再拘泥于黄金周，只要有吸引人的产品，他们会选择周末出行，甚至兴之所至，马上请几天年假就走。

2.3出行常态化

Web2.0时代是经济快速增长的时代，国人的度假消费越来越常态化。根据国际惯例，人均GDP达到3000美元，旅游形态开始向度假旅游升级，达到5000美元则开始进入成熟的度假经济时期。考虑到国人相对节俭的消费习惯以及近年来居住支出的大幅上涨，这一门槛会有所提高。但即便以人均GDP5000美元作为启动标准，2010年后，我国东部沿海地区——北京、山东、江苏、上海、浙江、福建、广东等省市都已超过人均GDP5000美元，进入到度假经济时代；而其中京沪两地的人均GDP已超过10000美元，迈入成熟的度假经济时代。中国旅游度假需求呈现持续增长态势，国家旅游局发布的《中国国内旅游抽样调查》显示，国内城镇居民旅游出行目的中休闲度假所占的比例，已经由2000年16.8%上升到2010年的25.0%，与观光游览和探亲访友呈三分天下格局。

中国国内旅游者出行目的的变化

旅游目的	观光游览	度假/休闲	探亲访友
2000年	39.9%	16.8%	26.3%
2010年	32.9%	25.0%	31.0%

旺盛的度假需求促进了旅游度假地的发育成长，一方面，传统的旅游度假地受到热捧，国家旅游局颁布的第一批12个国家级旅游度假区在旅游旺季几乎都是人满为患，海南等热门度假地更是一房难求；另一方面，旅游度假“蓝海”市场发育，长尾效应显著，这为区域级的小型旅游度假地成长提供了机遇。

2.4 渠道草根化

Web2.0时代，旅行者获取信息的手段极大增强，旅游度假地信息获取不再仅限于通过旅游出版物、旅游社和官方网站，而是更多地来自各种社会化网络。在Web2.0时代，旅游者、旅游区、旅行社、旅游管理部门都是信息的发布者。而获取信息的旅游者则更倾向于接受草根信息，如微博、驴友论坛、点评网站等。Web2.0时代，各省旅游局都开通了微博；世界知名的旅游目的地指南Lonely Planet也已开通新浪微博@LonelyPlanet，并已拥有10万粉丝；而旅游界的网络红人@行走40国，目前已经拥有近25万粉丝，成为草根意见领袖。Web2.0时代，传统旅游媒介的影响力大大削弱，草根意见则不断通过Web2.0平台聚合强化，并对旅游度假者的选择产生直接影响。

2.5 营销口碑化

Web2.0时代，用户发布与分享信息成为习惯。微博系统中有旅游者创建的海量信息，他们"随手拍、随手记、随手发"，与亲朋好友互动，与几亿网民共享旅游见闻、旅游感悟和旅游心得，直接将旅游目的地意向传播出去，间接达成了口碑营销的效果。而越来越多的旅游者出行前已习惯通过网络获取旅游目的地信息，他们虽然浏览主流门户网站，但更相信亲友和同事的推荐，甚至陌生人发布的点评信息也会成为考量的依据。Web2.0时代，一条微博、一句点评、一张图片，都可能成为旅游者选择目的地的关键要素。

3、Web2.0时代国家级旅游度假区规划设计新趋势

旅游度假区规划设计在我国起步较晚。1992年国务院批准建立12个国家级旅游度假区以来，20年间未见新增一例国家级旅游度假区。10年之前，有学者在分析了国内外旅游度假区发展历程后指出"随着度假旅游的快速发展和旅游度假区之间竞争的加剧，定位核心客源市场，形成度假区特色及满足游客多方面的需求，成为旅游度假区发展的新趋向，具体体现为主题性、文化性、生态性、景观性、休闲性五个方面[3]"。而今，旅游度假区又往前走过10年，在首批国家级旅游度假区提出20年之际，笔者尝试对旅游度假区发展新趋势和规划设计方向作进一步梳理。在Web2.0大潮下，当前旅游度假区规划设计思路的转变可概括为 "蓝海、原创、极化、混合、全时、多赢"等六大关键词。

3.1 蓝海

与20年前相比，旅游度假区已经从新生事物转向普及产品。原来以政府为主导，以满足人民群众需求的旅游度假区公益型开发模式，逐渐转向以市场为主导，以旅游产业带动区域综合发展的商业型开发模式。以Web2.0时代盛行的"蓝海战略"和"长尾理论"考量，传统旅游度假区开发着眼于"红海"，关注高等级旅游度假资源和规模化大众市场；在Web2.0时代则转变为开发广阔的"蓝海"，避开竞争激烈的"红海"，即挖掘不同层次的旅游度假需求，通过个性化、差异性的产品设计，满足广泛的旅游度假需求。1992年以来的20年，尽管12个国家级旅游度假区的数量没有变化，但省级旅游度假区发展迅猛，各种未定级的度假村、乡村度假设施、小型度假设施建设更是方兴未艾。以经济和人口总量较大的山东省为例，2000年省级旅游度假区已达到14家，2012年更已达到25家。

Web2.0时代，各种档次的旅游度假区建设都面临难得的机遇。高端旅游度假区可瞄准成熟度假市场需求，走精品化提升道路；中端旅游度假区可开发大众度假产品，立足本地市场，走规模化道路；而一般度假区可以开发特色度假产品，走差异化发展道路。总而言之，尽管中低等级的旅游度假产品知名度不高，利润不高，但其总体市场份额可以和高等级旅游度假产品的市场份额相比，甚至更大。而且中低等级的旅游度假区建设投入不大，建设限制因素较少，面向持续发育的国内旅游度假市场，具有广阔的发展空间。

3.2 原创

在强大的信息洪流推波助澜下，Web2.0时代的"眼球经济"比任何时代都要活跃。新兴旅游度假区要出奇制胜，才能吸引度假者眼球。这就要求旅游度假区的规划设计要发挥想象力，一个优秀的策划案能为旅游度假区增色，创新的规划或前卫的建筑甚至能成为旅游度假区的特殊吸引物。传统旅游度假区"以自然为基础，文化为灵魂"的规划思路，在Web2.0时代进化为"以自然和文化为本底，以创新体验为动力"的新模式。传统旅游度假区以夏威夷等热带海岛和圣托里尼岛等文化海岸为代表；而今，当自然吸引力和异文化吸引力已经无法吸引成熟度假消费者眼球的时候，迪拜的棕榈岛、新加坡的圣淘沙以标新立异的创意横空出世——迪拜以天马行空的人工岛规划，圣淘沙以独特的空中泳池设计赚足眼球，原创设计成为项目推广的强大动力，这与Web2.0时代的信息传播特征相吻合。

3.3 极化

传统旅游度假区专注于提供规模化、标准化的产品，而Web2.0时代需求的个性化趋势，促使旅游度假区产品呈现多极分化的特征。常规度假产品面临发展瓶颈，新型度假产品必须在某一方面做到极致，契合某一细分人群的度假需求，才能在激烈的市场竞争中立于不败之地。在高端市场，以悦容庄为代表的小型精品酒店（SLH, Small Luxury Hotels）获得长足发展，成为高端奢华度假的代名词。在传统中端市场，旅游度假区星级宾馆已经演变为酒店集群，国内知名旅游地产企业万达集团开发的长白山国际旅游度假区规划9个顶级度假酒店，首期一次性建成包含6家顶级酒店的度假集群，引入柏悦、威斯汀、凯悦、喜来登、假日等酒店管理集团，靠规模优势取胜；在低端平民市场，各旅游度假城市的商务快捷酒店如雨后春笋般蓬勃发展，如家、汉庭、七天、速八、锦江之星等后起之秀靠性价比优势牢牢站稳脚跟。

3.4 混合

Web2.0时代，用户交互创造新价值，达到1+1>2的效果。旅游度假区规划设计也呈现强烈的混合特征：通过多种度假产品组合，多种旅游业态组合、多种土地利用性质组合，满足度假者的个性化需求，创造全新的度假体验，进而促进物业增值。我国早期的旅游度假区设计强调功能分区，曾有学者对国内外旅游度假区土地利用情况进行了对比研究，提出了旅游度假区"带状、核式、双核式、多组团式[4]"等布局形态。而最新的旅游度假区理论，提出了诸如旅游综合体[5]、HOPSCA等混合性使用的新思路，模糊了传统度假区规划设计中各功能分区的界限。当前众多的旅游度假区建设项目中，往往通过度假设施与商业设施、文化设施、娱乐设施、旅游地产、养老地产、高尔夫球场、滑雪场的复合，达到旅游活动互相促进、区域景观互相提升，人气活力互相提振的效果。

3.5 全时

一般旅游度假区都具有明显的淡旺季，而Web2.0时代旅游度假者的"碎片化"特征，使得开发淡季旅游产品成为可能。越来越多的度假地开始做旅游低谷的文章，提出全年（旺季与淡季）、全天（白天与夜间）、全天候（室内与室外）等全时概念，以提高旅游度假设施利用效率，提高旅游度假地收益水平。以马来西亚云顶集团开发的云顶乐园为例，该旅游度假项目将冬季冰雪度假产品与夏季避暑度假产品相结合，规划上将冬季滑雪场与夏季高尔夫球场进行统一设计，极大地提高了旅游度假地的设施使用效率。

3.6 多赢

Web2.0时代 "去中心化"的草根特征带来了旅游度假区开发中多方利益博弈的新格局，这主要体现为两个方面：一是作为弱势群体的本地社区将获得更多的知情权和话语权，他们的权益得到更多的重视，越来越多的旅游度假区开发项目在居民拆迁安置方面采取更为人性化的措施，考虑解决居民再就业和社区产业转型的问题。二是公共利益得到更多关注和重视，公私博弈中公共利益能得到更好的保障，旅游度假区规划设计中服务于大众旅游者的空间和设施得到重视，而非完全向高端旅游者和高收益项目的倾斜。这些都从一定程度上促进了多方共赢和社会公平。

4、总结与展望

旅游度假区已经在中国走过20年，随着人均GDP水平的快速提高，我国即将踏入旅游度假区全面发展的新时代。与此同时，信息社会已踏上Web2.0快车道，更有学者提出了Web3.0概念，旅游度假区应紧跟时代潮流，主动调整规划设计思路。

[1] 金准. Web 2.0和旅游目的地营销系统. 旅游学刊, 2006.07, P11.

[2] 应丽君. Web 2.0冲击下我国旅游电子商务的新机会. 旅游学刊, 2007.05, P7-8.

[3] 周建明. 旅游度假区的发展趋势与规划特点. 国外城市规划, 2003.01, P25-29.

[4] 刘家明. 旅游度假区土地利用规划. 国外城市规划, 2000.03, P13-16.

[5] 陈琴, 李俊, 张述林. 旅游综合体开发——一种旅游房地产全新开发模式研究. 资源开发与市场, 2012. 04, P47-50.

New Trends for Tourist Resort Planning and Design in Web 2.0 Era

[Abstract] With the development of information technology (IT), we have entered into Web 2.0 era, whose typical product – blog has influenced the tourism activities unconsciously with its concept – "user generated content , taking the network as a platform for information sharing ". In Web 2.0 era, new characteristics come forth in tourism industry, such as the individualization of demand, fragmentation of time, popularization of traveling, grassroots channels, word of mouth marketing, and so on. In response, the planning and design idea for tourist resorts should change accordingly to adapt to the new trends in web 2.0 era. The change reflects in the following six new trends: blue ocean, original design, polarization, mix, full time and win-win.

Key Words: Web 2.0, tourist resort, planning and design, trend

Profile

Wang Binshan
Director of Beijing Tsinghua Urban Planning & Design Institute (THUPDI), Dept. of Tourism & Scenic Area Planning
Doctor, senior engineer, national registered urban planner

He was admitted to Tsinghua University in 1993, majoring in architecture, urban planning and landscape successively, and gotten his doctor's degree there. After graduation, he began to work in Beijing Tsinghua Urban Planning & Design Institute (THUPDI) and now he is the member of the council, the director of Dept. Of Tourism & Scenic Area Planning, and the deputy director of Landscape Research Center. With deep research and rich experience in the planning and design of scenic areas, tourism areas and tourism towns, he has participated in many key projects such as Mount Tai, Huangshan Mountain, Tibet, Beidaihe, Hainan International Tourism Island and so on. He is also the consultant for Wanda, Vanke, R&F, CR Land in tourism developments.

1. Introduction

The Information Technology Revolution is one of the greatest revolutions in the contemporary era. In Web 2.0 era, web models, such as blog, based on the concept of "user generated content, taking the network as a platform for information sharing", have established bridges between tourists as well as tourist and tourism destination. They have brought changes to the tourism activities unconsciously and influenced the planning and construction of the tourist resorts greatly. The developers should change their thoughts and adapt to the new trends in Web 2.0 era.

The term was brought forth by Dale Dougherty and Graig Cline in 2004 and greatly influenced the internet industry. Web 2.0 is based on the thoughts and theories such as Six Degrees of Separation, The Long Tail, Social Capital, Decentralization and so on. Major features of Web 2.0 include "user-generated content, co-creation, decentralization of the power of internet, using the wisdom of crowds to participate in the construction of the system, the rise of the long tail market, and the 'forever beta' service of the internet"[1]. "The new business pattern based on Web 2.0 will promote the reshuffling of tourism industry."[2]

2. New Characteristics of The Tourism in Web 2.0 Era

2.1 Individualization of Demand

The interactivity of Web 2.0 technology realized the mutual-communication between suppliers and customers, which makes the tourism products meet more requirements of the customers. By online booking and communicating, the customization of individual service is possible. With the increasing of individual demands, polarization occurs in the tourism market: traditional tourist resorts provide standard and popular tourism products, rely on famous e-commerce tourism websites, and follow the rules of economies of scale; while more and more small-scale tourist resorts appear to provide individualized and boutique products. They are focusing on the sub markets and following the rules of the experience economy. De Mao Tang Club, Huangshan and The Emperor Hotel Beijing are great reference for this type, because both of them have taken advantage of the location and local characteristics to provide individualized experience, and as a result found a place for themselves in the high-end tourism market.

2.2 Fragmentation of Time

Web 2.0 era is also the era for fragmented information. Traditional concentrated thinking model is gradually replaced by fragmented multi-way thinking model. And the fragmentation also appears in tourism activities. In Web 2.0 era, tourists are easy to be attracted by some experiences or just a hot topic, and then start a travel. Experiences such as "pack up and set off" and "feed the pigeons in Paris at weekends" become a cozy sign of life, and time is not the limitation for travel. People will not wait for golden week holidays but choose to travel at weekends or on annual vacation.

2.3 Popularization of Traveling

The economy grows fast in Web 2.0 era, and traveling becomes more and more popular in Chinese people. According to international practice, when per capita GDP reaches 3,000 US dollars, the sightseeing travel will upgrade to holiday travel; when it reaches 5,000 US dollars, the market will enter into the mature period of the holiday economy. In consideration of Chinese people's consumption habit of frugality and the increasing expenses in housing in recent years, the threshold will be higher. However, even with the per capita GDP 5,000 US dollars as the standard, the east coastal areas such as Beijing, Shandong, Jiangsu, Shanghai, Zhejiang, Fujian and Guangdong have entered into the holiday economy era after the year 2010 with a per capita GDP surpassing 5,000 US dollars. Among them, Beijing and Shanghai have first stepped into the mature holiday economy era with a per capita GDP surpassing 10,000 US dollars.

The tourism demands grow sustainably. According to the sample survey on China's Domestic Travel by National Tourism Administration of The People's Republic of China, the ratio of traveling for holiday rises from 16.8% (2000) to 25.0% (2010), which is equal to sightseeing tour and visit tour.

Changes of The Purpose for China's Domestic Travel

Purpose	Sightseeing	Holiday/Recreation	Visit
Year 2000	39.9%	16.8%	26.3%
Year 2010	32.9%	25.0%	31.0%

Strong demand drives the growth of tourist resorts. On one side, traditional tourist resorts are popular. The twelve first state-level tourist resorts are crowded in tourist seasons. For example, in the hot scenic area Hainan, it is hard to find an accommodation in the tourist seasons. On the other side, the Blue Ocean market grows with distinct "long tail effect" which benefits small-scale tourist resorts.

2.4 Grassroots Channels

In Web 2.0 era, tourists can get information by many ways. Besides the tourism publications, travel agents and official websites, there are also many social websites that provide travel information. Tourists, resort developers, travel agents tourism and the administration departments can broadcast the tourism information. And the receivers will be ,more inclined to draw information from grassroot channels such as blogs, BBS, comment websites and so on. Many provincial tourism bureaus have launched their official blogs, and Lonely Planet for the guide of world famous tourism destinations has its account on Sina Weibo and has already attracted about 100,000 followers. And the web celebrity "walk in 40 countries" has attracted about 250,000 followers, and thus has become the leader of grassroot tourists. In Web 2.0 era, the influence of traditional tourism media has weakened, in stead, the suggestions from common people have great increased and affected the tourists choices directly through the platforms provided by Web 2.0.

2.5 Word of Mouth Marketing

In Web 2.0 era, posting messages and sharing information become the habits of the internet users. Massive information can be found on weibo: tourists take photos during their tours, record and post the tour experiences on weibo and share them with families and friends as well as the internet users. They broadcast the experiences in the tourist destinations which benefits the word of mouth marketing. More and more tourists will get the information about the destinations from internet, and they will trust more of their friends, colleagues or even the comments of strangers'. In this era, a piece of blog, comment or picture may play a decisive role in making decision for a tourist destination.

3. New Trends for Tourist Resort Planning and Design in Web 2.0 Era

The planning and design for tourism resort started late in China. Since 1992 the State Council approved the establishment of 12 state-level tourist resorts, there have been none new project in the following twenty years. Ten years ago, some scholars analyzed the development history of overseas tourist resorts and pointed out that, "with the rapid development of tourism industry and the intensifying competition between tourist resorts, new trend of targeting major customers, forming self-characteristics and meeting multiple requirements has led the development of tourist resorts. It embodies in theme, culture, ecology, landscaping, recreation"3. Now, another ten years passed, and I am trying to further analyze the new trends in the development of tourist resorts as well as the direction for the planning and design of these areas. In Web 2.0 era, the keywords for the new design ideas can be summarize as "blue ocean, original design, polarization, mix, full time, and win-win".

3.1 Blue Ocean

Compare to twenty years ago, tourist resorts become more popular than new products. The public service based development model that supported by the government has changed to be market-oriented. New model of tourism driving the development of the whole area has become popular. According to the popular "blue ocean strategy" and "long tail theory" in Web 2.0 era, traditional tourism resort development focuses on "red ocean" and pays attention to high-class tourism resources and the scale of the market; now it changed to develop the vast "blue ocean" and try to meet different tourism requirements by providing individualized service and products. Though the number of state-level tourist resorts has not increased in the past twenty years since 1992, a great variety of provincial resorts, holiday villages, countryside tourism facilities and small resort facilities are developed fast. Take Shandong province as an example, there have been 14 province-level tourist resorts till 2000, which increases to 25 till 2012.

In Web 2.0 era, tourist resorts of all levels are facing new opportunities. High-end tourist resorts can improve themselves and make the mature market as target; mid-end ones can develop the popular tourism products based on local market and expand the scales gradually; for the other small-scale ones, characteristic products and different experiences are the new solutions. In a word, though mid- and low-end tourist resorts are not so famous and beneficial, they can win a great share of the tourism market even bigger than high-end ones. In addition, the investment for these mid- and low-end ones is smaller and the limitations for construction are less, they will have a further development in the future.

3.2 Original Design

With vast quantities of information, "attention economy" is more active in Web 2.0 era than in other eras. Newly built tourist resorts should present more surprises to draw people's attention. Thus the planning and design must be imaginative to help to enhance attraction with innovative planning or unique architectures. Traditional design idea is based on nature and culture, while new design should be also based on nature and culture, and be driven by innovative experience. Traditional tourist resorts are represented by the tropical islands and cultural beaches such as Hawaii islands and Santorini. And today, new destinations such as the artificial islands - the Palm Islands of Dubai, and Sentosa Island of Singapore with unique sky swimming pool, have won great popularity because of their innovative designs. Original design plays an important role to promote the tourism projects, which works in the same way the information spreads in Web 2.0 era.

3.3 Polarization

Traditional tourist resorts focus on providing large-scale and standardized products, while tourists of Web 2.0 era need more individualized service and products. Thus the tourism products gradually show the feature of multiple polarization. Conventional tourism products start to encounter bottlenecks, and the new-type ones must adapt to meet the specific requirements of some people and, then they can get the share in the competitive market. For the high-end market, small luxury hotels (SLH) represented by Banyan Tree have a significant development and become the name card of luxury resorts. In the mid-end market, star resort hotels are changed to hotel groups. For example, the famous tourism resort developer Wanda Group has set nine top resort hotels in Changbai Mountain such as Park Hyatt, Westin, Hyatt, Sheraton, Holiday Inn, and so on. In the low-end market, a great variety of business hotels spring up, namely, Home Inn, Hanting, 7 Days, Super 8 Hotel and Jinjiang Inn, have gained their own market shares by high cost performance.

3.4 Mix

In Web 2.0 era, user interaction creates new values and presents the effect of 1+1>2. The planning and design of the tourist resorts also shows the feature of mix: varied tourism products, programs and properties are combined together to meet individual requirements and present brand new tourism experiences. And as a result, it adds values to these properties. Traditional tourism resort design emphasizes the distribution of the functions. There have been ever scholars done researches to the tourism areas and summarized the conventional layouts to "belt, core, double core, and multi group"4. And according to the latest tourism resort theory, mix developments such as tourism complex5, and HOPSCA are introduced to break the boundaries between different functional areas. Today, in many tourism developments, tourism facilities are combined with commercial facilities, cultural facilities, entertainment facilities, tourism properties, health properties, golf course, ski field, etc. to enhance the popularity of each others.

3.5 Full Time

Generally, tourist resorts feature busy seasons and off seasons. However, the feature of "fragmentation" in Web 2.0 makes it possible to develop tourism products for so-called "off seasons". More and more tourist resorts begin to follow the concept of "Full Time" - full year (busy seasons and off seasons), full day (day and night) and all weather (interior and exterior). Off seasons are used to enhance the economic performance of the tourist resorts. For example, Genting Resort, developed by Malaysian Genting Group, has combined the programs of winter resorts and summer resorts together, integrating ski filed with golf court to raise the utilization rate of the tourism facilities.

3.6 Win-Win

"Decentralization" of Web 2.0 era brings new opportunities for different tourism developments, which embodies on two aspects: 1. Local residents get more rights and reasonable relocation compensation. The problems of re-employment and economic transition are carefully considered. 2. It pays more attention to the public benefits. Public facilities and resources are protected in the planning and construction. In this way, it will get a win-win development for many parties and safeguard social fairness.

4. Summary and Prospect

Tourism resorts have been developed for twenty years in China. With the fast increase of the per capita GDP, China is entering an era of all-around development for tourist resorts. At the same time, the information society has gotten on the Web 2.0 express. Some scholars even present the concept of "Web 3.0". Thus the tourist resorts should adapt to the new trends and change the ideas of planning and design.

(1) Jin Zhun. "Web 2.0 and Tourism Destination Marketing System", Tourism Tribune, 2006.07, P11.

(2) Ying Lijun. "New Chance for Tourism E-commerce in Web 2.0 Era", Tourism Tribune, 2007.05, P7-8.

(3) Zhou Jianming. "Development Trend and Planning Characteristics of the Tourism Resort", Urban Planning Overseas, 2003.01, P25-29.

(4) Liu Jiaming. "Land Planning in the Tourism Area", Urban Planning Overseas, 2000.03, P13-16.

(5) Chen Qin, Li Jun, Zhang Shulin, "Research on A New Model of Tourism Properties - Tourism Complex Development", Resource Development & Market, 2012.04, P47-50.

旅游度假区的发展与规划

广东省社会科学院旅游研究所总规划师、广东中建设计有限公司董事长
陈南江博士

一、中国已经进入度假产品时代

从1978年中国旅游产业化以来，旅游产品的发展经过了三个阶段：分别是以风景名胜区和文物古迹为主，延续历史并继承下来的观光产品阶段；1982年开始的以游乐园和主题公园为主的游乐产品阶段以及1992年开始的度假产品阶段。1992年国务院批准设立国家级旅游度假区，1996年国家旅游局确定以“休闲度假游”为年度主题，目前旅游市场也正从走马观花的“到此一游”旅游转向近距离的多次的休闲度假。相应的，也就从团队旅游转向散客自驾车旅游，而且旅游节奏逐渐放慢，时间观念淡化，游客可以自行掌握度假行为。进入21世纪以来，房地产与旅游度假走向紧密结合，度假村、度假区向度假社区快速跃进和提升，第二居所、投资性度假房产、疗养房产成为重要的房地产产品，度假产品开发进入了高潮。

各类度假产品分析

在进行度假产品开发时，首先必须认识到它的开发与传统的观光产品有很大的不同，即度假产品往往对资源不存在依赖性。度假产品分为资源依赖型与非资源依赖型两种。资源依赖型度假产品主要是温泉度假区和海滨度假区。非资源依赖型的度假产品开发在选址上存在较大的灵活性，其吸引旅游者的条件主要是优美的环境、丰富的设施、良好的服务，这三个方面是运用资金和管理即可以营造的。

根据地理环境的不同，度假区可分为滨水度假区、温泉度假区、山地度假区以及乡村度假区四类。

滨水度假区是以水环境为依托，包括海滨、湖泊、河流以及三者之中的岛屿。滨水度假区又可分为滨河型度假区、滨湖型度假区和滨海型度假区三种。其中滨河型度假区，由于河流通常有汛期，水位季节性变化明显，对于旅游建设影响较大。河上往往需要同时发展航运，加上水质原因，往往不适合开展水上游乐活动。滨湖型度假区由于湖泊水体的流动性相对较差，旅游开发对于水体质量的影响十分明显，因此必须高度重视污水的排放和处理。滨海型度假区是世界旅游度假区的主体，其特点是海洋的景观价值在度假区是次要的，主要是作为活动背景，为游客提供水面游乐和水下娱乐。目前滨水旅游度假区的数量在不断上升，一方面是不少水库转向旅游度假区；二是海滨沙滩被更多地开发出来；三是一些山谷被人为蓄水，形成旅游度假区。

温泉度假区是近几年国内度假区开发的热点，属于资源依托型的度假区，具有一定季节性和广泛的群众基础，重游率高。温泉度假区温泉的水温、水质和水量十分关键。目前国内温泉度假区基本上都是大众休闲温泉，缺乏高端温泉和疗养温泉，而且存在以下不足：第一，只有园林、别墅、酒店、温泉池，形式单调；第二，温泉虽有养生形式，但是指导性针对性不足；第三，温泉缺少文化的支撑，少量温泉有娱乐项目，但没有系统的文化主题；第四，没有明确的目标市场细分。针对这一现状，新开发温泉的突破口有三个，一是从园林温泉走向主题温泉和疗养温泉；二是要面向特定市场；三是提供特定的服务。

山地度假区具有很大季节性，是度假区开发难度较大的产品。山地地形对于开发具有较大影响，在山上选择合适的建设用地比较紧张。度假村的海拔高度、湿度等对于经营影响也很大，不同游客对山地度假的需求也不同。此外由于山地生态敏感，要高度注意生态保护。

国内的乡村度假区是由农家乐发展而来的，并依托古村落或有特色的小村落，分为原生乡村旅馆式和新建乡村度假村，往往规模较小。

度假区的选址要点

度假区的选址应根据其区位来判断，距离目标市场2小时车程以内的度假区，主要依托环境，城市近郊的甚至可以作为第一居所；距离目标市场2－4小时车程的度假区，主要依托资源；距离目标市场4小时车程以上的度假区，则必须依托高品位的资源或者优良的气候，如海南三亚。

度假旅游的特点

度假旅游更加注重休息和放松；度假旅游以近距离为主；度假旅游的目的地数量少；度假旅游者在一地停留的时间长，花费大；度假旅游者对旅游目的地导游的依赖程度轻；度假旅游呈现出家庭化特征；度假旅游的季节性强；度假旅游重游率高。

关于度假旅游区的误区

主要有两个：一是认为度假区就是在郊区的酒店，二是认为度假区就是园林、酒店与娱乐活动三者之和，目前国内度假区存在十分严重的同质化。度假旅游是高档次、高经济效益的旅游产品，度假区里发展的主要产品为度假酒店，另外还要发展会议酒店和适当的主题酒店。低层、低密度、低容积率、高绿化率是度假酒店的基本特色，即“三低一高”。对于传统的风景区而言，如何在政策和生态允许的前提下创造条件发展，让游客能够留下来，为度假区提高经济效益，是观光产品“二次开发”的重要课题。

目前度假区开发面临的问题

一是国务院的土地政策收紧，度假区项目与房地产开发相结合需要大量土地，土地征用难度增大；二是政府越来越重视农民的拆迁安置，不得采取强硬手段，必须进行协商；三是一些项目前期申报了自然保护区、风景名胜区等，规划管理约束较严，制约景区开发；四是一些企业抢占资源，而不是开发。

二、旅游度假区规划的基本逻辑

影响旅游度假区的基本地理因子共有九个，包括区位、气候、地形、地质地貌、动植物、水文、土壤、交通、人文。

地理环境对于旅游度假区的影响是综合性的。在进行旅游资源评价时，应该看到高级别资源有条件发展为高级别产品，一流的开发理念可以把二流的旅游资源开发为一流的旅游产品，但文物价值、生物价值、地质价值不等于旅游价值。

旅游产品竞争力主要是由区位（交通条件）、知名度、美誉度（口碑）构成的。其中口碑和知名度取决于三个方面：产品的特色，可以细分为环境特色、项目特色、文化特色；产品的经营营销策略，包括价格体系、形象策划与传播手段；提供的服务，即服务是否能让顾客满意。

旅游规划的任务不是仅仅安排项目布局和基础设施，如修路、建设大门等，让游客进入各个景点，或是有地方住。旅游规划应当形成旅游区特色、卖点和竞争力。旅游规划单位应当对投资者的规划经费负责，应当对投资者的巨额投资负责。一个好的规划应该能够为投资方赚钱、找钱和省钱。

一些规划单位编制的旅游规划成果之所以不适用，在于其规划工作路线不合适。旅游规划与城市规划的本质区别在于后者主要是将空间形态的安排作为行政管理的依据。旅游规划就远远不止如此，旅游规划应当包含旅游策划，它需要进行旅游市场问卷调查，需要进行周边旅游产品调研，需要进行认真的策划。如果没有做这三个方面的工作，规划的基础就十分薄弱，就难以遵循市场导向原则、特色原则和区域合作原则，难以编制出优秀的规划成果。因此判断一个旅游规划文本是否有价值，编制单位必须要明确目标市场是哪里、竞争对手是哪些、卖点在哪里以及游客是否会感兴趣等问题，否则规划就是失败的。

在进行旅游规划时首先要分析自身竞争力，分析周边同类产品和自身优势，同时还要以创新理念开发产品，即要努力超越传统思路，升级换代。对于新产品而言，要分析产品的发展阶段，避免跟风模仿，要积极创新。

三、旅游度假区的六大规划理念

旅游度假区的规划，在理念上应贯彻市场导向、强化特色、主题化发展、以人为本、生态优先以及强化操作等六个原则。

市场导向原则

市场是旅游规划研究的出发点和归宿。研究并满足游客的心理，是旅游企业成败的关键。市场经济就是竞争经济，旅游目的地必须正视和面对竞争，打造自己的核心竞争力。另外还要注意潜在的市场需求，游客潜在的需求在很大程度上需要旅游产品来激发。具体应做到以下三点：首先，要对目标市场进行科学定位。要明确不同族群游客所追求的利益，了解家庭生命周期与旅游喜好，灵活应用度假区市场预测的三种方法——宏观总量比例预测法、目标市场调查分析法以及类似项目比较预测法。其次，要致力于把握趋势。研究旅游业的发展趋势，研究地区经济和社会的发展，研究领先于群众的消费趋势并擅加运用于旅游开发。最后，要关注与旅游度假区关系密切的六大社会发展趋势：一是汽车时代来临，二是进入老龄化社会，三是对于青少年教育高度重视，四是投资第二住宅、居住郊区化，五是“亚健康”群体日益壮大，六是美容美体产业快速发展。这六个趋势，对应的是庞大的正在上升的细分旅游市场。

强化特色原则

旅游度假是卖方市场，没有特色就没有生命力。特色具有时间和空间的相对性，同时特色应当具有可操作性。旅游度假区的特色营造主要有四个方面，即环境特色、项目特色、文化特色与经营特色。

主题化发展

旅游度假区的主题化与文化有密切关系。除了极个别的以纯生态为主题的旅游度假区外，各类文化是度假区的主要选择主题。主题化是形成一个度假区文化特色的核心内容，它可以减少雷同。目前国内很多酒店和度假区都缺乏个性。主题可以是地方文化挖掘、历史资源的挖掘利用，也可以是外来文化的引进。

旅游度假区的主题化要注意四个问题：一是主题要具有市场吸引力；二是建设和经营必须到位，要充分展示该文化的各方面内涵；三是规模不大的度假区就不能多主题，企图“一网打尽”所有不同需求的客源；四是文化主题的表现手法不能单一，只局限于柱廊、诗碑、雕塑等静态陈列，要活化文化，使之生动和鲜活。如与湿地相关的文化除了人们熟知的生态文化外，还有与之相关的历史文化、饮食文化、商品文化、建筑文化等。

以人为本原则

旅游业不仅是一个经济学行业，更是一个心理学行业。研究并满足游客的心理，是旅游项目及旅游企业成败的关键，不同市场有不同的需求。总之，要在充分了解和满足游客需求的前提下，进行巧妙的策划。策划要注意以下几点：

1、尽量形成自己的卖点。例如领先（一定区域内第一个）、规模（一定区域内最大、最高、最多）、最好（一定区域内风景最美、环境最佳、服务最好、最有人文、最悠闲、最有参与性）、有独特的项目（最刺激、最好玩）、有独特而鲜明的主题以及面向特定市场。

2、遵循旅游策划的八条思路和途径：①移植，如日式温泉、普罗旺斯；②仿古，如唐式；③集锦，如世界温泉文化大观园；④形成鲜明主题；⑤面向专一市场；⑥反常规，如把房子盖到树上；⑦引入新技术，如“人造月亮”；⑧提出新概念，向房地产业学习。

如果是民俗文化类资源，进行旅游开发可以根据不同情况采取以下六种模式：集锦荟萃式、复古再现式、原地浓缩式、原生自然式、主题附会式以及串点连线式。

在功能区布局方面，要遵守五个原则：一是要符合旅游者行为需求和管理需求；二是功能分区相互间区分要明确，即静区与动区应当分离、高档消费区与低档消费区应当分离，私隐活动区与公众活动区应当分离；三是要符合生态化原则；四是要因地制宜，充分结合当地地形地貌等工程条件进行布局；五是游览线路要主次分明，尽量不让游客走回头路。

度假区的游乐活动，应尽量考虑游客的安全保护，做好安全提醒工作以及安全制度的安排，以减少事故的发生。如一些山谷型河流水位变化大，应留出泄洪道，注意建筑高程，大坝应对山洪有一定的预防和应对措施。河流附近的滨水建筑物可以底层架空处理。对于海滨度假区，应在地震发生前后密切注意海啸预警。其次，要考虑游乐活动的隐患。游泳、划船、潜水、拖曳伞、冲浪等水上活动随时可能造成游客的生命安全隐患。旅游度假区应当形成完善的安全救护管理制度。对天气不良、游客水性不佳、游客身体状况、超容量拥挤等情况向游客提供安全警示、救生设备、陪护员工，安排安全瞭望以及救生人员等。

生态优先原则

生态优先理念主要有两个重点：一是首先要创造良好的休闲度假生态环境；二是在经济与环境的矛盾中，以生态为优先。

强化操作原则

可操作性旅游规划要有产业联动的思想。一是与传统农业结合发展农业旅游；二是与传统林业结合发展林业旅游；三是与传统文化结合发展文化旅游；四是与乡村生活结合发展乡村旅游；五是与房地产业结合发展产权度假旅游。评价一个旅游规划的标准有三个，分别是体系的完整性、内容的科学性和创新性、实施的可操作性。目前我国度假区开发的要点主要是：一是尽量整体开发或大地段开发，避免小地块切割开发，确保风格的统一和管理的方便；二是确保低层、低密度、低容积率、高绿化率；三是加强环境整治，协调风格和色彩；四是将高中低档市场在地段上分开发展；五是除了乡村度假产品之外，一般的度假产品尽量避免村庄对于度假村的干扰。旅游度假区与居民的空间关系主要有两种模式：一种是欧洲模式，其特点是度假区与社区融合在一起；另一种是亚洲模式，特点是度假区与社区分开发展。我国度假区至今建设较好的，基本上是不自觉地采取了亚洲模式，那些在村落基础上发展而来的度假区或者按照城镇规划的度假区，基本上都是不太成功的。在新的拆迁安置政策背景下，越来越多的度假区将不得不走向欧洲模式，如何在亚洲走向成功是一大挑战。

■结语

从景区到度假区，再到当今的度假社区，旅游产品开发与规划设计与时俱进。如果没有高瞻远瞩、可操作性强的规划，就难有好的作品，难有好的效益，各个投资单位与规划机构应细分市场，研究市场，以扎实的工作作风打造满足市场需求、激发市场需求的成功作品。

Development & Planning of Tourist Resort

Chief Planner of Guangdong Provincial Academy of Social Science Tourism Research Institution, President of Guangzhou Zhongjian Design Co. Ltd., Dr. Chen Nanjiang

1. China has entered the era of vacation products

Since 1978, China's tourism products have undergone three stages: inherited tourism product stage featured scenic spots and cultural relics; amusement product stage from 1982 featured amusement park and theme park and vacation product stage form 1992. In 1992, the State Council approved the establishment of national holiday resort. In 1996, National Tourism Administration confirmed "leisure travel" as annual theme. The former snapshot-style travel has turned into leisure in a closer distance. Accordingly, team travel turns to self-driving travel, the pace of travel slows down gradually and the concept of time fades, thus visitors can enjoy the travel on their own. Since the beginning of 21st century, real estate and tourist resort walked into a closer relation. Second home, investment vacation real estate and recuperate property have become important real estate products. The development of vacation products reaches a climax.

Analysis on Various Vacation Products

Before developing vacation products, we must first recognize they are very different from the traditional tourism product, for they do not rely on resources. Vacation products are divided into two types: resource-dependent and non-resource-dependent. Hot spring resort and beach resort belong to the former one. Regarding to the later one, it has greater flexibility in the site conditions to attract tourists. Beautiful environment, rich facilities and good service, these three factors can be created by money and achieved by management.

Depending on the geographical environment, the resort can be divided into waterfront resort, hot spring resort, mountain resort and village holiday resort.

Waterfront resort relies on water, including coast, lakes, rivers and islands. It can be divided into riverfront resort, lakeside resort and coastal resort. Regarding to riverfront resort, it is not suitable to develop water recreation activities for the flood season and significant seasonal water level changes of the river, which has great impact on tourism development and it often requires the simultaneous development of river navigation.

Regarding to lakeside resort, it must attach great importance to the discharge of sewage and treatment for the slow flow of the water and caused damages. Coastal resort is the main part of world tourism resort, which is characterized by the landscape value of ocean in the resort area is secondary, but act as the background that provides visitor with water recreation and underwater entertainment. The number of waterfront tourist resort is rising. The reasons are: firstly, many reservoirs have turned to tourist resorts; secondly, sandy beach is developed in a further scale; thirdly, some of the valleys are artificially filled with water to become tourism resort.

Hot spring resort in recent years is the hot spot in resort development, which belongs to the resource-dependent type, boasts with seasonality, broad mass base and high rate of re-travel. Water temperature, water quality and quantity are essential elements of this kind. Domestic hot spring resorts are basically mass leisure spa but lack of high-end spa with the following deficiencies. Firstly, the monotonous form with only garden, villa, hotel and hot spring pool; secondly, health maintenance with insufficient guidance; thirdly, only entertainment projects but no systematic cultural theme; fourthly, no clear target market segments. In response to this situation, there are three ways to make breakthrough: one, moving toward to theme hot spring and spa but not garden hot spring; two, targeting to specific markets; three, providing specific service.

Mountain resort is hard to develop for its seasonality. Mountainous terrain limits the site. In addition, the altitude and the humidity of the resort really matter to the business and different visitors have different requirements on mountain resort. In addition, we should pay great attention to ecological protection for the mountain ecosystem is very sensitive.

Domestic village holiday resort is developed from farmhouse that relying on an ancient village or a small featured village. It always small in scale and is divides into native country-style resort and new village resort.

The Key to Site Selection

Site selection should be determined by zone bit. The resort within 2 hours' drive away from the target market mainly relying on the environment, suburban can be used as the first residence; the one within 2-4 hour's drive away from the target market mainly relying on resources; the one who keeps more than 4 hour's drive away from the target market should be able to rely on high grade resources and good climate, like Sanya, Hainan.

Features of Vacation Tour

Vacation tour pays more attention to rest and relaxation and prefers destination at close range. Number of vacation destination is small. Vacation tourists usually spend a lot of time and money in one place, and they do not depend so much on guides. Vacation tour tends to present family characteristics. And it boasts strong seasonality and high rate of re-travel.

Misunderstanding on Tourist Resort

There are two main misunderstanding on tourist resorts: one is that tourist resort is hotel in the suburbs, the other is that tourist resort is the sum of garden, hotel and entertainment facilities. Currently there is very serious homogenization in domestic resorts. Vacation tour is a high-grade and high-cost tourism product. In addition to conference hotel and proper theme hotels, the main product in the resort is resort hotel, whose features are low-rise, low density, low plot ratio and high greening ratio. For traditional scenic spot, how to create conditions the policy and ecology allow to make visitors stay to improve the economic efficiency is an important topic of tourism products' secondary development.

Current problems faced by resort development

Firstly is the State Council's land policy tightening, the combination development of resort project and real estate requires a lot of land, which increased the difficulty of land acquisition. Secondly is that government pays more and more attention on resettlement of farmers, tough measures must be taken in consultation. Thirdly, a number of projects declare nature reserves, scenic spots, etc. in an earlier stage, stringent management constraints the development. Fourthly, some enterprises just seize the resources rather than developing.

2. Basic logic of tourist resort planning

There are nine geographical factors the tourist resort matters, including location, climate, topography, geological features, flora and fauna, hydrology, soil, traffic and culture.

Geographical environment has comprehensive influence on tourist resort. We should see the conditions to develop high-level products when evaluating tourism resources. With first-rate development concept, second-rate tourism resources would be developed into first-rate tourism product, but cultural relic value, biological value and geology value are not equal to travel value.

Location (traffic condition), popularity and reputation compose the competitiveness of tourism product. Reputation and popularity depend on three aspects: the characteristics of the products which can be subdivided into environmental characteristics, project characteristics and cultural characteristics; marketing strategy including price system, image planning and the means of communication; the service provided, i.e., whether the service satisfied the customers.

Tourism planning is not only about to arrange the project layout and infrastructure such as building roads and gates, guiding visitors to scenic spots and providing them a place to live, but to become the feature, selling points and the competiveness of the tourist resort. Planning and design companies shall be responsible for the planning funds and huge investments of the investors. A good plan should be able to make money, find money and save money for investors.

The reason why some plans are not applicable is because the planning route is not appropriate. The essential difference between tourism planning and urban planning is that the latter regards the arrangement of spatial form as the basis of administration. Tourism planning is far more than that, which shall include scheme requiring market survey, product research and further serious planning. Without doing these, the weak start of planning is difficult to follow the market-oriented principal, characteristic principal and regional cooperation principal, and not likely to work out excellent outcome in the end. Therefore, to judge a valuable tourism planning text, the planning company should be understand where the target market is, who the competitors are, what the selling point is and whether the tourists would be interested or not, otherwise, it is a failure one.

Before the planning, we should first analyze our own competitiveness, advantages and the surrounding similar products, to develop products with innovative ideas, i.e., exceeding traditional ideas, upgrading and updating. For new products, their development stage should be analyzed so as to avoid copycat and create actively.

3. Six planning concepts of tourism resort

Six principles should be carried out during the tourist resort area planning, they are market orientation, strengthening characteristics, development of thematization, people-orientation, eco-priority and strengthening operation.

Market orientation

The market is the starting point and destination for tourism planning. Researching and meeting the demand of visitors' psychological demands is the key to the success of tourism enterprises. The market economy is the competition economy. Tourist destinations must face up to the competition and build their own competitiveness. The focus should be as follows: First of all, to figure out target market, to define the interest pursued by different groups of visitors, to understand the family life cycle and travel preferences and to apply three market forecasting methods flexibly—macro total ratio predicted method, target market survey investigation and analysis and method of comparative analysis on similar projects. Second, to study the development of tourism trends, regional economic and society, to research advanced consumer trends and applied them to tourism development. Finally, pay attention to the six social trends that related closely to tourism resort: the advent of automobile age, the coming of aging time, great importance attached to adolescent education, the second house investment and residential suburbanization, the growing of sub-health group and the fast development of beauty and body industry. These six trends correspond to the large rising segments of tourism market.

Strengthening Characteristics

The tourist resort is a seller's market, no features, no vitality. Features are the relative time and space, and features should be operable. Features to create tourist resort have four main aspects, namely, environmental characteristics, project characteristics, cultural characteristics and operating characteristics.

Development of thematization

Thematization has a close relation with culture. In addition to the rare pure ecological theme, all kinds of cultures are the main theme of tourism resort. Thematization is the core content to form the cultural characteristics of a resort, which can reduce similarity. At present, many domestic and resorts are lack of personality. The theme could be the excavation of local culture, utilization of historic resources and the introduction of foreign culture.

There are four issues that the thematization should pay attention to: first, it must have market attractiveness; second, construction and operation must be in place to fully show the various cultural connotations; three, small resort should not use multi-themes, in attempt to catch all the different needs of tourists; four, technique of expression of the cultural theme shouldn't confined to a static display of colonnade, monument, sculpture, etc., but to activate culture, make it vivid and alive. In addition to the well-known ecological culture that related to wet land, there are historical culture, food culture, commodity culture and architecture culture as well.

People-orientation

Tourism is a psychology industry rather than economics industry. To research and satisfy tourists' psychological needs is the key to the success of tourism projects and tourism enterprises. Different markets have different needs. In short, they have to plan skillfully under the premise of fully understanding and meeting the needs of tourists. Following points are what should pay attention to:

1. Try to form they own selling points, such as taking the lead (within a certain area), scale (within a certain region, the highest, the maximum), the best (the most beautiful scenery, environment, service, culture and participatory), unique projects(the most exciting and the finniest), distinctive theme for a specific market.

2. Follow eight travel planning ideas and ways: ① transplantation, such as Japanese-style spa, Provence; ② antique, e.g., Tang style; ③ a collection of choice specimens, like the World Hot Spring Culture Grand View Garden; ④forming bright theme; ⑤facing to a specific market; ⑥be unconventional, as building houses on the trees; ⑦introducing new techniques, such as artificial moon; ⑧presenting new concepts and learning from real estate industry.

Regarding to folk culture resources, tourism development can take the following six patterns based on different situations: highlight assembling, vintage reappearing and autochthonous concentration.

Regarding to functional area layout, it has to comply with five principles: one, to meet the needs and management needs of tourist behavior; two, function division should be clear between static area and dynamic area, high-end consumer area and low consumption, private area and public area; three, to comply with ecological principles; four, act local; five, making a distinction between the important and the lesser one on touring route, don't let the tourists go backward.

The resort's recreational activities should consider the safety of tourists, arranging security system in place to reduce accidents. Some of the valley's river levels should set aside spillway for the variation of water level. Pay attention to building elevation, the dam should be strong enough to prevent torrential flood. The bottom of riverside building can be built on stilts. Regarding to seaside resort, tsunami warning should be paid close attention to before and after the earthquake. In the second place, we must consider the hidden dangers of recreational activities. Swimming, boating, diving, parasailing, surfing and other water activities may bring potential safety hazard at any time. Therefore, perfect aid management system should be formed and improved, providing safety warning and lifesaving appliance for tourists on various situations, arranging observation space and lifeguard, etc.

Eco-priority

There are two main points in this part: we must first create a favorable ecological environment for vacation relaxation; regarding to the contradiction of the economy and the environment, environment got a priority.

Strengthening Operation

Operational tourism planning should be led by the thought of industry linkage. 1. Tourism links with traditional agriculture to develop agricultural tourism. 2. Tourism links with traditional forestry to develop forest tourism. 3. Tourism combines with traditional culture to develop cultural tourism. 4. Tourism combines with country life to develop country tourism. 5. Tourism links with real estate industry to develop property tourism. There are three standards to evaluate the tourism planning: complete system, scientific and innovative content, and operability. At present, the features of the tourism development for China include: 1. Overall development and large-area development are applied to avoid small and separate development as well as ensure consistent style and centralized management. 2. Ensuring low buildings, low density, low plot ratio and high greening ratio. 3. Enhancing environment improvement and coordinating styles and colors. 4. Separating the high-, mid- and low-end markets by locations. 5. Except the country tourism, other tourism products should try to get rid of the disturbance from surrounding countries. Thus the spatial relationship between tourist resorts and the residents can be summarized into two patterns: one is the European pattern which features the combination of the resorts and the surrounding neighborhoods. The other is the Asian pattern which has separated the resorts from the neighborhoods. Most of our well-developed tourist resorts usually apply the Asian pattern unconsciously, while the resorts based on countries or following the country-style planning are not so popular. With new relocation policy, more and more tourist resorts have to apply or transfer to the European pattern. It will be a big challenge for Asian market.

Conclusion

From scenic spot to the resort area and then to today's resort communities, development and planning design of tourism products keep pace with the times. It there is no visionary and operable plan, it is hard to make good works and economic returns. Every investor and planning agency should subdivide and research the market, with solid work style to make successful projects that meet and stimulate market demand.

EXOTIC WITH COMPLETE FACILITIES

Lion Lake Arab Conference Hotel, Qingyuan

异域风情 设施完善—— 清远狮子湖阿拉伯会议酒店

项目地点： 中国广东省清远市
建筑设计： 华森建筑与工程设计顾问有限公司
总占地面积： 120 600 m^2
建筑面积： 93 152 m^2

Location: Qingyuan, Guangdong, China
Architectural Deisgn: Hussen Architectural & Engineering Designing Consultant Ltd.
Total Land Area: 120,600 m^2
Floor Area: 93,152 m^2

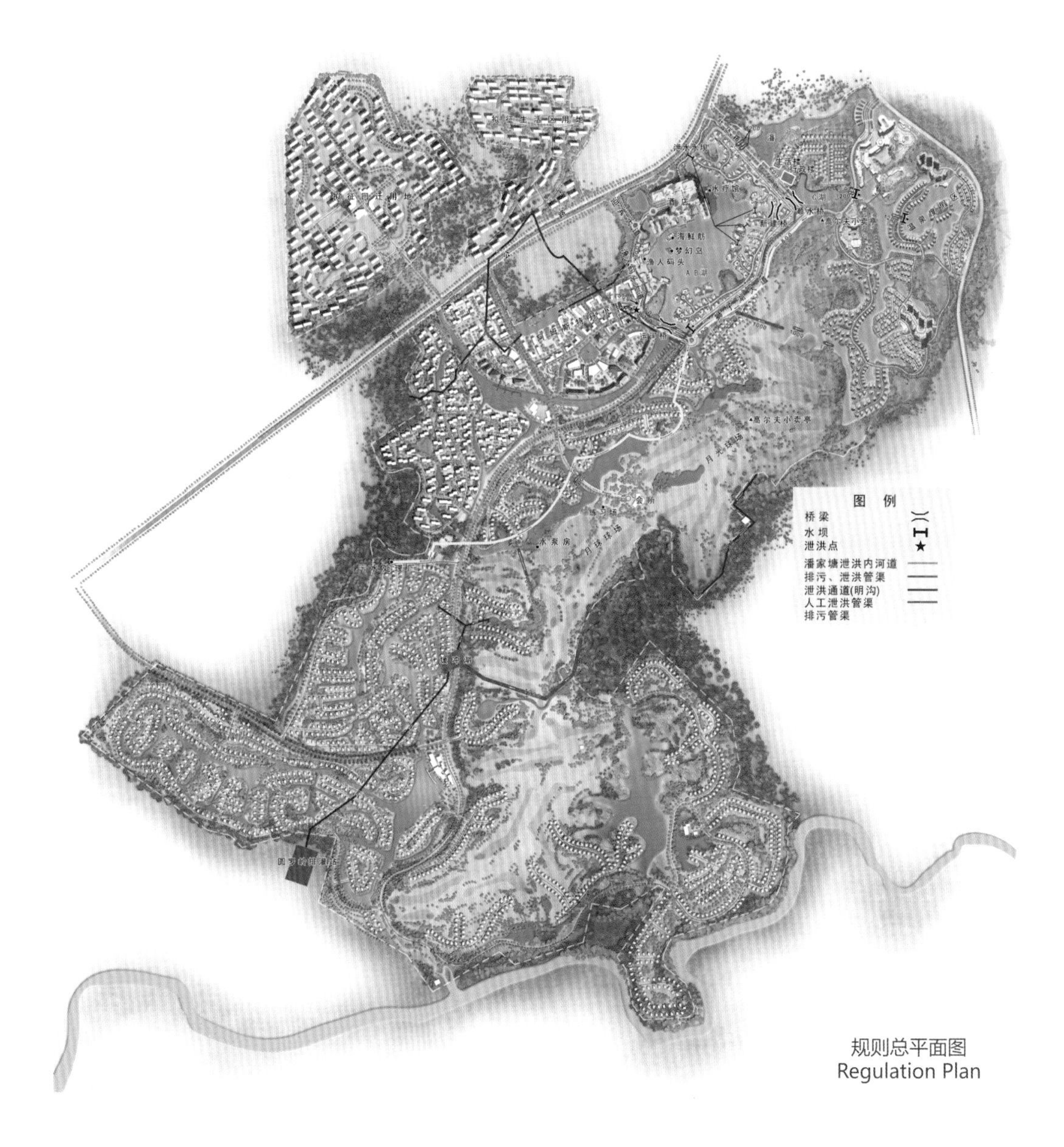

规则总平面图
Regulation Plan

项目概况

清远市狮子湖阿拉伯会议酒店项目按五星级标准开发。酒店项目地块位于清远市横荷街道狮子湖畔，地块位置优越，南临107国道仅5分钟车程，北靠清远至三水公路，交通方便，环境幽雅。

规划布局

狮子湖主题酒店地块呈三角形状，总用地面积为120 600 m^2，初步规划由酒店及相关服务设施构成，总建筑面积为93 152 m^2。酒店拥有3万m^2的大型会议中心、351间五星级酒店客房、带私人游艇码头的总统套房、10间顶级SPA房、8间海鲜房等规模与设施，集度假、商务会议、宴会与餐饮、健身娱乐中心、水疗中心、主题别墅等多种功能于一身，既有天方夜谭般的梦幻，又有金碧辉煌的皇家风范。周边还设有中西式餐饮、会议中心、娱乐及篮球、网球等各种球类运动场、游泳池等项目，配备完善。

建筑设计

该酒店以阿拉伯建筑符号为特色，以空中花园的建筑形态、一千零一夜的人物传说雕塑和壁画打造一个具有异域风情的商务会议酒店。

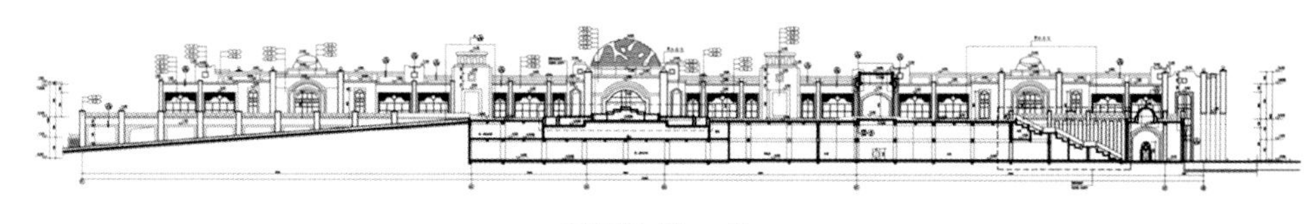

立面图 Elevation

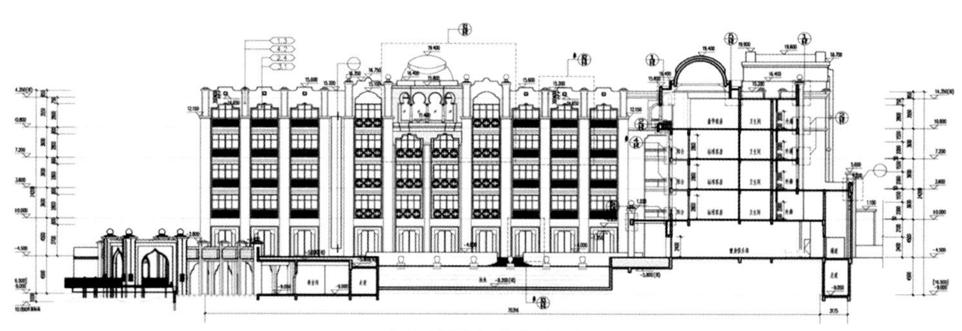

1-1 剖面图 1-1 Section

该项目的智能化系统包含综合布线系统、计算机网络系统、背景音乐及公共广播系统、机房工程等15个子系统。

景观设计

狮子湖社区配有晨岛、茶岛、夜岛、梦岛、花岛等几大主题岛，每一个岛临水临花临草，风光旖旎；而一望无际、绿草如茵的国际锦标级高尔夫球场，也有被火红丹霞石点缀其中的月球球场、平湖如镜点缀其中的月光球场二个主题，景色怡人。酒店位于风水绝佳的狮子湖山峦，景观优美，可以浏览狮子湖全貌，美丽的倩影倒映在一片风清水柔的粼粼湖光之中，身临其境，不用远足仿佛便到了异国他乡。

Profile

Developed according to 5-star standard, the project is located beside Lion Lake in Henghe Street of Qingyuan wih superior geographic position. It is only 5 minutes' drive from the 107 National Highway and adjacent to Sanshui Highway with convenient traffic condition and beautiful environment.

Planning Layout

The hotel is of triangle shape with total land area of 120,600 m^2. It is made up of the hotel and the relative facilities with total floor area of 93,152 m^2. The hotel contains a large conference center of 30,000 m^2, 351 guest rooms, the president suite with private marina, 10 top SPA rooms, 8 seafood rooms, etc., collecting multiple functions of resort, commercial conference, banquet, catering, sports, entertainment, SPA, theme villas, etc. It has both the feature of fantasy and the style of royalty. The facilities around are complete, such as Chinese and western catering, conference center, entertainment, sports courts for basketball and tennis, swimming pool, etc.

Architectural Design

The hotel features the Arabian symbol, the shape of hanging garden and the statues and frescos of the characters from Arabian Nights, creating an exotic hotel for commercial conference.

The Intelligent System of the project includes Premises Distributed System, Computer Network System, Background Music and Public Broadcasting System, Computer Room Project, etc. There are altogether 15 subsidiary systems.

Landscape Design

There are several theme islands in the community of Lion Lake: Dawn Island, Tea Island, Night Island, Dream Island, and Flower Island. Every island is adjacent to water and flower with beautiful scenery. The vast and green international golf court also has two sections of different themes. The hotel is located among the mountains of Lion Lake with wonderful landscapes, where the full view of the lake can be seen.

以从都项目为例阐述国内旅游度假区设计要点

华森建筑与工程设计顾问有限公司执行总建筑师 郭智敏

■ 作者简介

郭智敏
华森建筑与工程设计顾问有限公司执行总建筑师
教授级高级建筑师
国家一级注册建筑师

1988年7月毕业于天津大学建筑系，1988~1997年就职于建设部建筑设计院，1997年至今就职于华森建筑与工程设计顾问有限公司，任公司的执行总建筑师。2009年被深圳建筑协会评为深圳十佳提名青年建筑师称号。

从概念上说，旅游度假区是指旅游资源集中，环境优美，具有一定规模和游览条件，旅游功能相对完整独立，为游憩休闲、修学、健身、康体等目的而设计、经营的，能够提供相当旅游度假设施和服务的旅游目的地整体。

无论是国内还是国外，旅游度假区的概念都起源于少数统治者消费闲暇时间的一种需要，如国外最早出现的为了满足执政官需要而建立的公共浴室和相应的旅店配套设施，国内则主要体现为皇家园林与私家园林，如河北承德的避暑山庄、北京的颐和园等皇家园林，以及苏州、无锡等地的私家园林。直至社会财产和权力逐渐分化至民众的近现代，才使大众休闲度假在东西方日益活跃，体现在国内则是始于20世纪90年代，以1992年国务院批准建立的12个国家旅游度假区为标志，我国的大众化度假旅游产品才开始启动。经过20年的发展，国内已形成为数不少的旅游度假区，成功例子不少，但其中也产生不少问题，如发展依托不得力、功能设置不齐全、个性特色不明显、缺乏持久吸引力等。

基于我国人口基数大，旅游资源丰富，且度假旅游正成为发展速度最快的旅游产品，旅游度假区有着巨大的发展前景，其规划设计还有很大的改进空间，设计笔者借鉴与国外著名旅游建筑设计机构合作规划设计的旅游度假区项目，结合自主设计的休闲度假项目，阐述对国内旅游度假区设计趋势的一些观点，希望为该类型建筑的规划设计带来积极有用的建议。

从都（原规划名称侨鑫养生谷）项目，由侨鑫房地产公司开发，规划中借鉴了国外成熟的旅游度假区设计理念和本土特色，结合温泉水疗、高尔夫、商务会议和高端房地产开发四大功能为一体，为典型的走高端路线的旅游度假区。

项目概况

本项目位于从化市东部，距广州市区约8公里的从化温泉镇附近。规划区坐落于流溪河支流杨梅坑河河谷原地带，地貌类型为低山、丘陵、河谷平原，面积各占1/3，西北东三面环山，南面为流溪河，具备良好的景观资源和充沛的温泉资源。除体育公园占地120公顷外，规划总占地面积为33公顷，建筑面积为9万m^2。

规划布局

从都的规划布局以皇家狩猎场为主概念，强调尊贵的皇家气息，因而强调建筑类型分布主次排列有序，另一方面，擅用群山环绕的地形条件也决定了建筑布局的具体形态自由而有机。

贯穿地块中心的主干道将项目划分成东西两个地块，两个地块都环绕各自主题布局。东地块依山而起，以白金五星级酒店、国际会议中心为中心，环绕客房翼、别墅客房、泳池、水疗中心等形成以商务会议为功能、以温泉水疗为特色的酒店建筑群；西侧地块依托27洞高尔夫球场，以尊贵奢华的高尔夫会所为中心，环绕以不同等级的别墅群，形成以高尔夫为主题的建筑群；贵宾院落式酒店位于园林式酒店用地西北角的坡地上，为整个项目的制高点，北向靠山，南向是一览无余的高尔夫球场，景观资源极佳，成为地块另一种意义上的最为尊贵的焦点。

功能内容

客房

• 酒店：90间客房，3层与4层高的六星级酒店。每个房间可以尽享绿带中的三种环境中之一种：山陵、高尔夫球场、溪流。

• 高尔夫贵宾别墅：134 m^2的套间。所有别墅均受到私密的入室门保护，像一座白天的凉亭，为高尔夫贵宾提供独占的感觉。每栋别墅都有一个放松身心的室外温泉亭。

• 山地别墅：10栋别墅，每栋560 m^2包括四间卧室。其中两栋是可以分用的套房，每户都可以看到高尔夫球场的景色。

• 高尔夫别墅：10栋411 m^2的别墅提供高度私密性的花园以及高尔夫球穴区的迷人景色。每户别墅有4间卧室，其中至少1户备有分用式的套房。

• 花园别墅：14栋186 m^2的套房。每户有1间卧室及1座水疗的亭阁。

• 果园别墅：6栋186 m^2的别墅，类似于花园别墅，位置于果园的景色中。

• 总统别墅：这栋别墅是项目范围内唯一可以开车到达的。堪称奢华缩影的别墅面积地占有2 800 m^2，位于最高点依山而立，再加上最高级的防御措施，传递出一种与世隔绝的独占感。这栋别墅配有厨房、泳池和娱乐室等等的休闲设施。更有一位全职管家24小时全天候命。

高尔夫会所

这座8 000 m^2的会所紧邻27洞高尔夫球场以

及5栋私密的高尔夫别墅。高尔夫球场和会所将成为珠江三角洲地区乃至国内最知名的，将能够举行高尔夫国际竞标赛。

会议中心

会议中心位于酒店大堂层下一层，内有大型宴会厅、会议室和联合国式的报告厅。会议中心将有自己单独的上下客区，并且对外开放，这也将是广州北部唯一的国际会议中心，可以接待大约500人的国际会议。会议中心将主要用于从化市政府接待、高级商务活动和大型会议的举行等。

水疗馆

1700㎡的温泉水疗馆能欣赏到部分的高尔夫球场景观同时将让酒店客人以及专程来享受水疗的当地客人沉溺于奢华享受中。

建筑设计

项目以中国皇家园林为雏形，以现代、古典相结合的手法将建筑群统一在一致的建筑语言下。建筑形式以现代的手法演绎中式的神韵，院落式的建筑布局是其最大特点，建筑物与院墙有机结合，共同融入周边环境；富层次的造型做出视觉趣味，从形式、尺度、比例、组织等方面寻求中式建筑的精髓，以现代的材料和结构体系构筑出真正的传统中式感觉。

赋予不同功能、不同等级的建筑以一定差异化外观设计，使整个建筑群既统一又等级分明，隐含中国皇家建筑的等级秩序。

高尔夫会所，作为中心建筑，采用的是高等级的皇家建筑风格，在周边的单檐歇山建筑群中以十字形的重檐歇山顶成为视觉的中心，并以斗栱、人字栱等华丽的细节元素强调其尊贵感，是我们在休闲建筑中融入中式皇家建筑气息的具有突破意义的尝试；白金五星级酒店的中心建筑则相对低调，虽然体量大，且功能复杂而形成建筑群，仅在入口处以重檐歇山顶建筑和塔楼的形式进行强调；位于制高点的总统别墅则以高耸的基座、重檐叠瓦的群体形象、华丽丰富的细部装饰从建筑群中脱颖而出。

本项目的规划设计经过与政府部门、境外的室内外设计公司、甲方、酒店管理公司和旅游管理公司等机构的长期磨合过程，并经过多方对国内外成功案例的研究分析和借鉴，才逐渐成熟并最终建成，结合本项目和所参与的其他旅游度假区项目设计，笔者尝试总结出国内旅游度假区的设计要点：

一、主题明确——也就是特色为先。所谓主题，指的是度假区发展的主要理念或核心内容，能形成或强化度假区特色，增强度假区的竞争优势，满足度假区核心客源市场的休闲度假需求。以从都为例，满足高端客户需求的皇家特色是其最大特点。随着社会经济发展，财产的积累有所区别，必然产生高消费的需求群体，养生谷项目紧抓此类人群，并将该主题发挥到极致，从而形成了广州乃至珠三角地区最有特色的旅游度假区之一。

二、综合模式——也就是多功能的混合。传统的度假目的主要是保健疗养，现代度假目的则逐渐扩大，除传统的健康消费外，亲情回归、社会交往、素质提升、会议商务、消费闲暇等也成为度假的目的，单一功能的度假区很容易引起审美疲劳，甚至很容易为其他新兴度假区所取代。复合的功能布局则能大大加强吸引力，并会产生相辅相成的作用。从都依托地段和地形优势有温泉酒店和高尔夫两大复合功能，且两者紧密联系，能同时满足旅客运动、休闲的需求；另外区内还有国际会议中心，能满足高端会议要求，并能以度假酒店满足其住宿要求。

三、景观利用——环境、资源依托是度假区选址必不可少的，就建筑规划和设计而言，一方面应尽量利用原有自然景观，同时要重视生态环境的保护，尽量减少旅游活动所带来的负面影响；另一方面积极创造建筑周边的“微景观”，让游客能处处观景，让身心得到最大的放松，而舒适的院落空间的营造是其中最重要的手段，能使私密性和观景需求间得到平衡。

四、差异化——即提供各种不同类型的房型选择，且加大差距。首先，高端产品的引入有利于提升旅游区的整体价值；传统度假游客大多以两人或三人家庭为基本单元，而伴随功能的多样化、人们消费水平的差异化，对房型的要求会增多，如家族出游可能更倾向于合院式的客房，公司培训、高端会议等就需要高管专属的VIP套房甚至总统别墅等。

五、文化诉求——随着生活水平的改善，人们必然会对文化诉求提出更高的要求，尤其是对旅游产业而言，追根溯源，无论东西方，从旅游度假区产生开始就与文化需求密不可分，文化是度假区的灵魂，是度假区能够存在与发展的源泉，是度假区形成特色的主要组成部分。而现阶段国内旅游度假区的文化体现主要集中在国外文化表达和本土文化表达两方面，前者因为游览者猎奇的心态和人们对发达国家生活的向往而在产业发展的前一阶段大放异彩，全国各地涌现出不同的外国主题度假区。而随着国家逐渐强大，民族自我认同感的提高，本土文化的诉求肯定会日益高涨。

我有幸主持了从都这个中式风格建筑群俱乐部的方案设计和施工图设计的全过程，向我们的历史建筑文化学习了很多，也体会了我们历史文化的博大与精深。对度假区而言，注重本国、本土文化的表述肯定是未来规划设计的重要趋势，我们希望以我们的学习与探索为旅游度假类建筑的设计与发展尽绵薄之力。

Presentation of Main Designing Points in Domestic Resorts——Taking Congdu Project as An Example

Guo Zhimin, Chief Executive Architect of Huasen Architectural & Engineering Design Consultant Ltd.

Profile:

Guo Zhimin,
Chief Executive Architect in Huasen Architectural and Engineering Design Consultant Ltd.
Professorial Senior Architect
First-grade national registered architect

Graduated from Architecture Department of Tianjin University in July 1988
Worked in Architecture Design Institute in Ministry of Construction in 1988 - 1997
Working as Chief Executive Architect in Huasen Architectural and Engineering Design Consultant Ltd. since 1997
Rewarded as Shenzhen Top 10 Nomination of Young Architects by Shenzhen Architectural Association in 2009

■ 从都项目建筑主要参加人员：

方案设计人 工程总负责人：郭智敏
从都酒店设计参加人：夏韬、吕飞、曾耀松等
从都俱乐部设计参加人：曾耀松、于源、吴凡、郁萍、史旭等
贵宾院落式酒店:郁萍、吴凡、于源、陈铂等
别墅客房：汤文健、施广德等

■ The Major Participants in Congdu Project:

Project Designing, chief supervisor: Guo Zhimin
Congdu hotel designing: Xia Tao, Lv Fei, Zeng Yaosong
Congdu club designing: Zeng Yaosong, Yu Yuan, Wu Fan, Yu Ping, Shi xu
VIP courtyard-style hotel designing: Yu Ping, Wu Fan, Yu Yuan, Chen Bo
Villa guestroom designing: Tang Wenjian, Shi guangde

Conceptually, resort refers to a tourist destination and establishment having intensive tourism resources and graceful surroundings, with a fair scale and tour condition, quite all-inclusive and self-contained tourism functions. Designed and run for relaxation and recreation, study, health building and protection etc, it is able to offer facilities and services of tourism.

Either at home or abroad, the concept of resort is originated from the leisure need of the minority rulers. For example, the earliest public bathhouse and relative hotel facilities overseas was set for archons; at home it was more easily to be found as royal and private gardens, like Royal Gardens of Summer Villa in Chengde, Hebei, of Summer Palace in Beijing etc, private gardens in Suzhou, Wuxi etc. Public tourism at leisure became popular in east and west when it came to the stage of modernization that social property and power had separated and come into the hands of common people. Domestically it began from 1990s, and the approval by the State Council in 1992 of building 12 national resorts marked the prelude of popularized vacation tourism. With 20 years' development, quite a few resorts have been established at home, from which not only successful cases but also failures and problems could be found, like weak development supports, incomplete functions in set-up, indistinctive features, lacking everlasting attraction and etc.

Considering that China has a large population and abundant tourism resources; moreover, its leisure tourism is becoming the fastest developed tourism product with great development prospect and large improvement space in its programming and designing, the author would like to present a few opinions about the designing trend of domestic resorts, by learning from some leisure tourism projects planned and designed by famous architectural design institutions abroad and from some projects of own design, in the hope that it will bring some advices to the programming and designing of such sort of architectures.

Congdu Project (Qiaoxin Health Valley as its original planned name) is exploited by Qiaoxin Real Estate Company. It learns from the mature design ideas and local characteristic overseas in the programming, and integrates fours functions into one: Spa, Golf, Commercial Conference and Exploitation of High-End Real Estates, with the self-definition as typical high-end resort.

Profile

The project locates in the east of Conghua City, closing to Conghua Hot Spring Town which is 80 000 m to Guangzhou downtown. The planned region lies in Yangmei Kenghe Valley, the branch of Liuxi River, with the geomorphologic types of low mountain, hill and valley plain, 1/3 of each in the total space. It is surrounded on three sides (north, ease, and west) by mountains; the south is Liuxi River. It is rich in landscape resources and hot spring resources. Aside from 1,200,000 m^2 of Sports Park, the total area is 330,000 m^2, the architectural area is 90,000 m^2.

Planning Layout

Congdu planning layout takes Royal Hunting Ground as its central concept, and lays the emphasis of royal nobility, resulting in the orderly arrangement for primary and secondary. On the other hand, the terrain condition of being surrounded by mountains requires the freedom and harmony of the concrete modality in the construction layout.

The main road through the region center divides the projects into two areas, the east and the west, both of them lays out under their own theme. Built on mountains, the east area arranges the guestrooms flanks, villa guestrooms, swimming pools, spa centers etc, encircling the center--- Platinum 5-Star Hotel and International Convention Center, which shapes itself to be an architectural complex with functions of commercial conferences and with feature of hot spring and spa. The west area is designed based on the 27-holes Golf Course and is made the noble and luxurious Golf Club as its center. It is an architectural complex surrounded by different grades of villas with the golf theme. The VIP Courtyard-Style Hotel is located in a sloping field in the north west of Garden-Style Hotel. It is the commanding point of the whole project. In addition, it has a mountain to the north and a golf course with an unobstructed view to the south. The excellent landscape resource makes it the most distinguished focus in the whole region in another sense.

Functions

Guestrooms

• Hotel: 90 guestrooms, 6-star hotel of 3-floor and 4-floors. One of three sorts of environment could be fully viewed in every room: Hills, Golf Course and Steam.

- Golf VIP Villas: suits of 134 m^2 each. All villas have the privacy protection in each door, like an arbor in the daytime, offering the sense of exclusiveness to golf VIP. Every villa has an outdoor hot spring pavilion for relaxation.
- Mountain Villas: 4 bed rooms and 560m^2each, 10 villas. Two of them are the suits able to be separated, and both of them allow catching sight of the golf course.
- Golf Villas: 411 m^2 and 4 bed rooms each, 10 villas. Each villa has at least one suit that could be separated. The villas have the garden with high privacy and the glamorous golf putting green.
- Garden Villas: 14 suits of 1,866 m^2, a bed room, a pavilion for spa each.
- Orchard Villas: 6 villas of 186 m^2 each, similar to Garden Villas, located in the beauty of Orchard.
- Presidential Villa: It is the only villa in the Project that could reach by car. Rated as the epitome of luxury with a floor area of 28,000 m^2, it is constructed in the apex based on the mountain. Moreover, it has the top-grade security, which shows its exclusiveness sense for seclusion. The villa is equipped with kitchen, swimming pool, and recreation facilities like recreation rooms etc. Furthermore, a full-time housekeeper is standing by 24 hours.

Golf Club

The 80,000 m^2 club lies next to 27-holes Golf Course and 5 private Golf Villas. The Course and Club will turn to be the most famous ones in Pearl River Delta region even in the whole country. It will be available to hold Golf International Tournaments.

Conference Center

Conference Center, in the lower floor of the hotel lobby, has large Banquet Hall, Council Chamber and UN-style Lecture Hall. The detached load/unload area in Conference Center will be open to public, and it will be the only international conference center in north Guangzhou that could hold international conferences with a capacity of about 500 attendees. Conference Center will be used mainly for Conghua governmental reception, high-level business activities, large conferences and etc.

Spa

Segmental view of Golf Course from the 1,700 m^2 Spa brings more enjoyments to the hotel guests, and the local guests who come all the way to enjoy the spa.

Architectural Design

Taking Chinese imperial gardens as its embryo, the Project integrates the modern and classical architecture styles into one concordant complex. It employs the modern style in expressing Chinese essence, which is its architecture type. The courtyard-style construction layout turns out to be its greatest feature. An ingenious integration of structure and courtyard wall blends both into the surroundings. The multilayer style brings out the pleasure in visual sense. It seeks the essence of Chinese architecture from style, size, proportion, organization and etc, embodies the real traditional Chinese sense with the modern materials and structure system.

The designs of the architecture appearance vary with their functions and grades, concordantly but hierarchically, implying the hierarchy in Chinese royal architecture.

Golf Club, the central building, adopts the high-level royal architecture style. The cross-shaped multiple-eaves Xieshan roof as a visual center is circled with a complex of single-eave Xieshan roofs, using gorgeous elements like bracket set, gable arch etc. to emphasize the dignity, which is our effort in the breakthrough to melt the vein of Chinese royal architecture into leisure architecture. The central building of Platinum 5-star Hotel is kind of low-key comparatively. In the architecture complex of large scale and multiple functions, the only emphasis lies on the entrance by the multiple-eaves Xieshan roof and tower. The Presidential Villa in the commanding point is distinguished itself from the complex for lofty foundations, the multi-eave and shingled image, luxuriant decoration in details.

The planning and designing of the Project was getting mature and finally completed under the analysis and study of successful cases at home and abroad, and under a long period of adjustment with government apartments, domestic and overseas design companies, Party A, hotel management companies and tourism management companies etc. Learning from the planning and designing of the Project and other projects participated in, the author attempts to summarize the main designing points in domestic resort:

1. Clear-cut Theme——Putting characteristics as the main priority. So-called theme refers to the main idea or core in the resort development, which could become or reinforce the characteristics, could strengthen its competitive advantage, and could meet the leisure need from the core tourist market. Take Congdu for instance, the royal characteristic of meeting the demand of its high-end customers is the most distinguished feature. The development of social economy and the gap of property accumulation will certainly lead to a need of high consumption in the consumer group. The Project concentrates on this group of people and emphasizes the theme greatly, accordingly turns out to be one of the most distinctive resorts in Guangzhou even in Pearl River Delta.

2. Aggregative Mode——Mixing up the multiple functions. The traditional purpose of vacation mainly aims to healthcare and recuperation, while the modern vacation purpose is enriching. Apart from traditional health consumption, family relations, social association, quality improvement, commercial conference and leisure consumption etc have become the vacation purpose too. Compared with single-function resorts which are boring aesthetically soon to the guests and are easily replaced by other newly rising resorts, resorts with complex functions are much more attractive and self-supplementary. Based on the advantage of terrain, Congdu has two main complex functions: hot spring hotel and golf. They are connected to each other closely and able to meet sport and leisure demands at the same time. In addition, the International Conference Center could meet the demand of high-end conferences and vacation hotel could meet the need of accommodations additionally.

3. Landscape Exploitation——Environment and resource are essential when considering the resort site. With reference to construction planning and designing, on one hand, the original nature landscapes should be made fully use if possible, while the protection of ecological environment should be paid attention to, and every effort should be taken to lower the negative effect caused by tourism. On the other hand, it is agreeable to build micro-landscapes around the architecture so that the tourists' sightseeing everywhere becomes real and they could get the great relaxation for their bodies and souls. It is the most significant technique to create a comfortable courtyard space, balancing the privacy and visual demand.

4. Differentiation——Providing different choices of room types, and enlarging their difference. First of all, the draw-into of high-end products will be helpful to improve the value of the whole resort. Most of the traditional type tourists are grouped in the unit of 2-people family or 3-people family. With the diversification of functions and the differentiation of consumption levels, the requirements about room types will be much more. Family tour might prefer courtyard-type guestroom, and corporate training, high-end conference will be held in VIP suits or Presidential Villa which are set for senior executives.

5. Cultural Request——With the improvement of living standard, people will consequentially raise a higher request for culture. Especially in terms of tourism industry, no matter in east or west, if traced to the sources, resorts has been associated inseparably with cultural request since the very beginning. As the soul of resorts and the major part in resorts' characteristic, culture is the source that supports the existence and development of resorts. At present the culture of domestic resorts is reflected mainly in the expression of overseas culture and of native culture. The former was signalized in the first stage of industry development for visitors' curiosity and the adoration of living in developed countries, various exotic resorts correspondingly sprang up all around the country. Along with the growing of national power and the enhancement of nationality identification sense, the appeal for native culture will be definitely growing high day by day.

It was our great honor having taken part in the planning and designing of Congdu Project, a club of Chinese-style architecture complex. We have learnt a lot from our historical architecture culture, realized the sophistication and extensiveness of our historical culture. In terms of resorts, the attention to mainland and native culture definitely will become the significant trend of planning and designing in the future. We hope we could devote ourselves to the designing and development of resort architectures by our learning and quest.

LANDSCAPE AS SETTINGS FOR EDUCATION AND EXPERIENCE

Qinhuangdao Botanic Garden

融科普及体验于一体的“背景”设计—— 秦皇岛植物园

项目地点：中国河北省秦皇岛
景观设计：北京土人城市规划设计有限公司
主创设计师：俞孔坚
占地面积：265 000 m^2

Location: Qinuangdao City, Hebei, China
Landscape Architect: Turenscape (Beijing Turen Design Institute)
Principal designer: Kongjian Yu
Size: 265,000 m^2

项目概况

项目位于河北省东岸著名的旅游城市——秦皇岛，占地面积约265 000 m²，昔日用作苗圃，有六家小型残破工厂盘踞于此，垃圾遍地，人皆病之。植物园的方案随之提出，将与串联山海的汤河沿岸的绿色走廊相连。

规划布局与景观设计

设计目标：设计主要有三个目标：营造一个公共开放空间，供当地居民自由进出参观游览；打造一个旅游胜地，进一步增强秦皇岛作为旅游名城的吸引力；作为植物展示、环境知识普及的工具，增强人们对周围景观环境的保护意识。

景观设计的主要理念是恢复景观作为“背景”的含义，融科普及景观体验于一体。现存的乔木与景观元素通过精心安排，与新的设计完美融合。场地上现有的成熟树木包括枣树、紫藤、金钟柏、宝塔树（山茶树）及树苗都被保留下来。苗圃中的主道也保留了下来，与两侧的树木一起，融入新的设计。该设计形成了一系列的“背景”——景区，用不同的方式向人们展示植物的魅力。这些景区设计的灵感多来自于当地景观。该园主要包括以下景观节点：

入口区：入口区主要由本地树种——银杏树阵围合而成，地面由当地传统的材料碎石及黑色砖块铺砌，给人以亲切感。砖墙则界定了入口的范围。

草药园：黑色砖墙围合成四个庭院；红色钢制横梁制成的屏幕则构成了连接封闭庭院的视觉通道，给人以现代感。庭院中种植着各式中草药。花岗岩石板为人们提供了歇息的地方，并与绿植并排，形成现代化的模式。

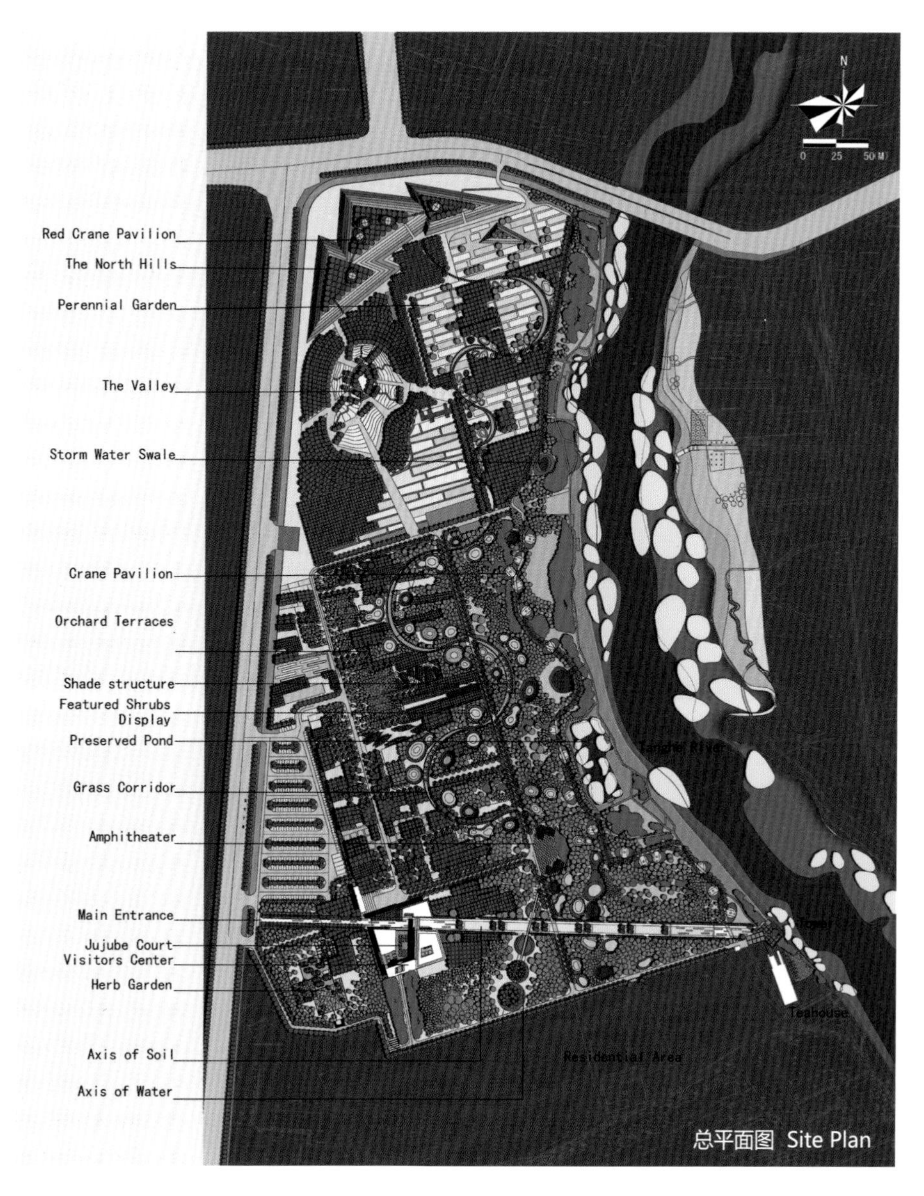

总平面图 Site Plan

枣树林：庭院围绕一片枣树林而建。这里建有下沉式花园，供人们在炎炎夏日坐享阴凉和果实。同样传统的黑砖材料再次用作铺装，构成了亲切的景观氛围。

土壤轴：该轴是入口轴在南北方向的延伸，沿与本地圆柏展开。轴中心是一块倾斜的频谱板，展示了土壤从岩石、沙粒、黏土到沃土的转变过程。同时，不同植物群落的精心布置，也向人们逐一展示了植物的“频谱”。

万花径：多年生花卉和本地草种，在该区极为常见。这些植物沿漫步道生长，形成了贯穿园区的主要步道。沿着木栈道，种植着多年生植物及草种。座椅及遮阳设计中融入藤本植物，与步道融为一体。人们可以零距离地接触植物，亲身感受植物的质感。

果林区：为了适应北方的温带气候，平地和坡道上种植了大片的秦皇岛本地果园，如梨园、苹果园、桃园、杏园、核桃园等，均沿地形呈阶梯状布置。果园地面铺装由灌木及花卉构成，在果树变绿之前为果林点缀斑斓色彩。其他的细节设计，如立面的微妙变化、色彩明亮的纤化玻璃的使用均改善了单调的景观效果。

山谷：设计师设计了一座人造山谷将5 m高的假山截为两半，露出由不同岩石形成的地质层，展示了当地著名的山地景观。本地植物群落遍布山坡、山谷。山顶架起几座桥梁，供人们通行并体验跨越山谷的精彩。

丘陵：在公园北端，原来的垃圾堆被改造成一座座小山，形成了丘陵地带，在寒冷的冬天阻挡了凛冽的北风，从而为植物和游客提供了微气候。

仙鹤亭：遮阳亭的设计灵感来自于水面常见的鹤，亭子上攀爬着藤本植物。多数亭子都是白色的，而位于山顶的五座亭子则设计成醒目的红色，用来吸引远处的游客。

通过采用当代设计策略，秦皇岛植物园将为人们提供连续不断的景观体验，是游客能在游园过程中学习到有关植物和环境的知识。该园功能多样，既是旅游胜地，又是环境、植物知识的教育基地，同时也是当地居民日常散步、游玩的开放空间。本园区区弹丸之地，短短两年营建，故不敢言聚物其丰，然汲燕山岚壑之精华，披渤海风露之灵气，更浸五千年华夏之人文，虽草木而文化，纵土砾而精神。园今告成，为港城蓬勃发展之见证，有科普教化、审美启智之功德，故铭文以志。

Profile

The project is located at the Qinhuangdao City, a tourism city at the east coast of Hebei Province of North China's. Totally 265,000 m^2 (about 65.5 acre) in area, the site is a former tree nursery with some existing seedlings, mature trees and roads, and six small factories that went bankrupted a few years earlier which make part of this site become a garbage dump with degraded environment. A botanic garden was proposed to transform this site, which will also connect the green corridor along the Tanghe River that connects the sea with the mountain.

Landscape Planning and Design

Three objectives are set for the design: (1)A public open space allowing local communities to visit freely;(2)A tourism attraction that can increase the capacity of the city as a overcrowded tourism city; and (3)An educational facility allowing people to learning about the native plants and environment and to be aware of their native landscape.

The main concept is to recover the meaning of landscape as "scenes" or "settings" that allow the plants to be displayed and visitors to play, allowing people to learn while experiencing the pleasant landscape. Existing trees and landscape elements are carefully evaluated and integrated into the new design: Existing mature trees on site including Jujube (Zizyphus jujuba), Purple vine (Wisteria sinensis) ,Arborvitae (Platycladus orientalis), Pagoda Trees (Sorphor Japonica) are kept and integrated into the new design, so are many existing seedlings; Major paths of the former nursery are kept because they are lined up with mature trees, which are also integrated into new design. A series of settings are created to display plants in various ways. Most of the setting is created at the inspirations of the vernacular landscape seen in the region. Some highlights of the park include:

The entrance: The entrance is composed of allies of Gingko trees native to this region, paved with gravels and black bricks, which are the traditional

local material that gives a sense of vernacular. Brick walls are used to define the entrance to the park.

Herb gardens: The black brick walls are used to create four courtyards; screens made of steel beams in red color are used to create visual access between the enclosed courtyards, which create a contemporary sense. Herbs used as Chinese medicine are gown in these courtyards. Granit slates are used for seating and are lined up with planters to create a contemporary pattern.

Jujube court: the courtyard was built around a grove of existing Jujube trees, and a sunken garden was made for people to sit and enjoy the canopy in the summer and abundant fruits from the tress. Again, the traditional material of black bricks is used for the paving to create a sense of vernacular landscape.

The axis of soil: it is the north–south extension of the entrance axis, lined with densely planted the native Chinese Juniper (Juniperus chinensis), and in the center of the axis is a gradient spectrum of various medium types changing from course rock, to sand, loam and fertile soil. Accordingly, dative plants communities are planted which created a spectrum of vegetation that change gradually as visitor walk into the park.

The perennial corridor: Perennial flowers and native grasses, which are seen with high diversity in species in these area, are grown along the meandering corridor that run across the park is one of the major routes for visitors to enjoy the park by foot. Following the board walk, perennial and especially native grasses are grown along the board walk. Seats and shadow casting structure that also support climbing plants are integrated into the design of the corridor. Grasses and plants are planted close to the board so that visitor can feel the texture of the plant, creating a moving through experience.

Orchard fields: As adaptation to the north temperate climate, the regional vernacular landscape in Qinhuangdao area is characterized with fruit orchards on the plain and at slopes. In this section of the park, Peach, Apples, Pear, Almonds, Walnuts, etc. are grown in terraces that are leveled based on the existing topography. Shrubs and flowers are used as the undercover of the fruit orchard, which will blossom before the fruit tress turn into green. Detail design elements, such as subtle difference of elevation, the use of bright colored fiber glass, are used to dramatize the otherwise monotonous landscape.

The valley: an artificial valley is designed that cuts through a man made mound (maximum 5 meters high) to display the geological stratum made of different kind of rocks. It is a depiction of the local mountainous landscape. Associated plant communities are planted in the valley and on the mound. Bridges are built on the top of the mound that allow people to have an unique experience of walking above the valley.

The hills: At the north end of the park, a series of geometric hills are built

using the garbage dump from the site, that screen of the cold wind in the winter from the north so that a comfortable microclimate is created for the plants and visitors in the park.
The Crane Pavilions: Inspired by the cranes which are common seen in the river corridor nearby, several pavilions are designed as shadow casting structures that will also support climbing plants. Normally they are in white color, but the five on the top of the hills are in red with intention to attract attention of visitors who can see from far way.

Using contemporary design strategies, the Qinhuangdao Botanic Garden is designed as an experiential landscape that created different settings or "scenes" sequentially, allowing visitors to have unique experiences in different setting during the process of walking and exploring through the park and learning about the plants and their environment at the same time. It is a park that serves multiple functions as a tourism attraction, an environmental and botanical education facility, and as an open space for daily use of the local communities. After two years since its opening to the public, we proudly say that it was a great success in fulfilling all these objectives.

SURROUNDED BY MOUNTAINS AND LAKE, JUST GROW IN NATURE

Thousand Island Lake Runhe Resort Hotel

背山面湖 长于自然的精妙布局—— 千岛湖润和度假酒店

项目地点：中国浙江省淳安县
规划设计：上海秉仁建筑设计有限公司
占地面积：16 661 m²
总建筑面积：53 984 m²
建筑密度：15.9%
容 积 率：0.39
绿 化 率：63.2%

Location: Chun'an County, Zhejiang, China
Planning & Design: DDB International Shanghai
Site Area: 16,661 m²
Total Floor Area: 53,984 m²
Building Density: 15.9%
Plot Ratio: 0.39
Greening Ratio: 63.2%

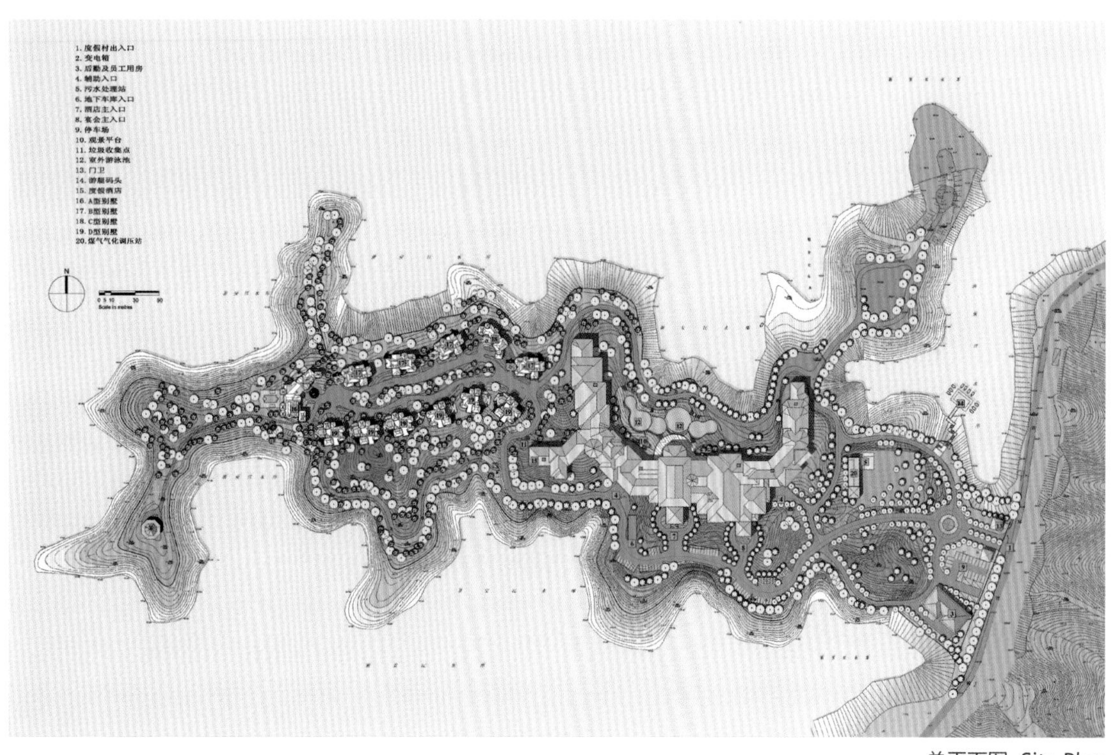

总平面图 Site Plan

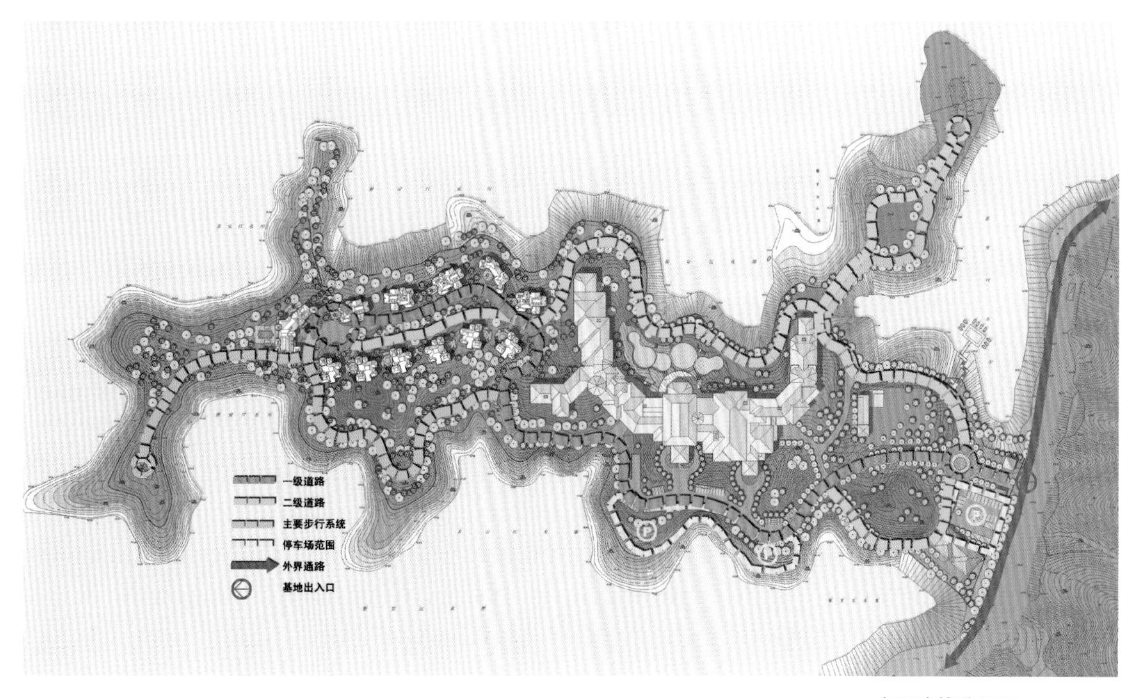

交通流线分析图
Traffic Network Analysis Drawing

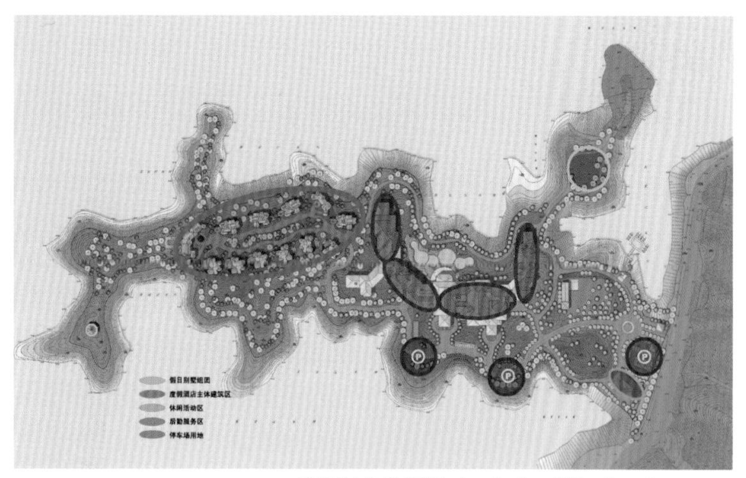

空间结构分析图 Analysis of Planing Structure

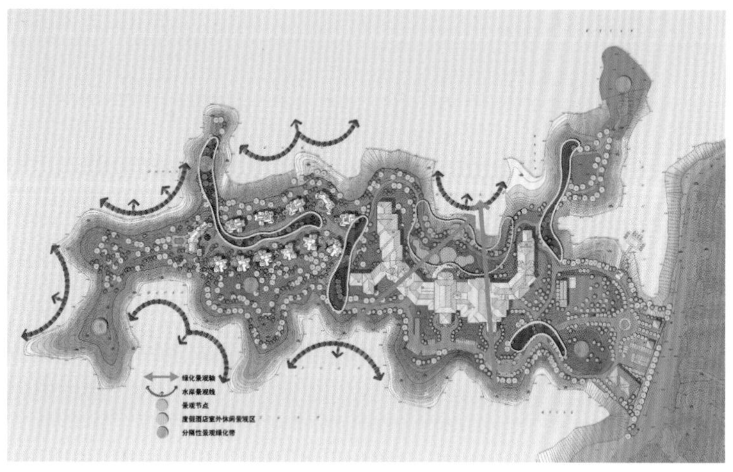

景观绿化分析图 Landscape Green Analysis Drawing

项目概况

本项目所在基地位于千岛湖中——东西向的半岛——梦姑岛之上。该岛南、西、北三面环湖，东面接临直通千岛湖镇的城市道路，使项目建成后有便捷的交通和完善的配套，同时项目本身也可以为淳安县创造一个良好的城市休闲环境。项目建成后背山面湖，尤其适合休闲度假。

规划布局

为了更好地适应基地的环境特点，对基地入口、道路、建筑群的布局和景观规划做出了精心的考虑，力求使建筑“自然生长”于环境之中。

入口：基地东面与城市道路接壤，利用这里的平坦地势，设置了度假村的出入口，在基地的入口处，排布了网球场及体量较小的健身中心，并且将建筑物局部下沉，使其掩映在葱葱绿荫中。这种局部方式不仅使入口处的用地得到充分的利用，同时还保持入口处良好的景观和开阔的视野。

道路：基地中的道路规划顺应地势开辟，主要道路沿基地高势布局，既合理地组织了基地内的二级道路，也合理组织了建筑物排布。基地中道路关系清晰明了，主次分明。规划道路中充分考虑了利用本基地内的景观资源，沿湖布置景观步道，连接了规划中的多个景观节点，使自然景观和人工景观有机地联系了起来，创造了一个供人休闲散步的场所和立体的生态系统。

建筑空间布局：基地中的度假酒店和假日别墅在空间排布上分区明确。酒店则布置在基地的中部，利用了中部坡度相对平缓、用地相对宽裕的位置，布置了主入口，这样不仅使入口用地比较宽裕，同时减少了土方量。酒店位置的排布还兼顾公共性和私密性的要求，因此在空间上结合道路设计，将假日别墅区排布在基地的西面，使该区域相对独立，不被其他闲杂车辆及人员穿越，便于安全管理，

并且该地块三面环水，使每幢假日别墅都能具有良好的水面景观。

建筑设计

天际线控制：建筑群尽量契合原有的地形铺开布局，同时又通过对建筑群层数的控制来压低建筑的高度，度假酒店除中心部分5层以外，其他基本控制在4层，别墅则控制在2~3层，以此来使建筑群和山体更好地结合起来。不仅如此，建筑单体也根据地形变化而高低错落，并且结合山地地形采用了退台的方式，这样不仅能使建筑能更好地依山而筑，同时从湖面上看时也能使建筑轮廓和原有岛屿的天际线保持一致。

度假区别墅区建筑单体设计说明：设计从两方面来确定设计原则，一方面考虑如何减少对千岛湖自然景观的破坏，使建筑融合成为新景观的一部分；另一方面是如何利用千岛湖自然景观，使自然景观不仅仅是别墅的优美背景，更融合成为生活中所有感官愉悦的源泉。

户型设计

房型设计中，突出“景观、度假、朝向”三大主题。入口玄关、各厅室、厨浴私人空间，以及庭院园林等室外空间的排布，皆以借景湖光山色为第一要素。营造起居室、餐厅、早餐室以及厨房之间的通敞，庭院空间的共享，一卧一卫的配比，多角度的景观窗式，满目美景以及阳光的露台等。

景观设计

在景观布置上充分考虑了景观的原生性，利用原有的自然景观，加以修整并且结合地形布置泳池、小品、亲水平台等人工景观，使人工景观和自然景观充分融合，满足休闲度假的使用功能。在植物配置上尽可能保护原有绿化品种，加以梳理，同时根据季节变化配载一些其他色彩的植物种类，形成物种多种多样的生态群落。

Profile

Located on Dream Girl Island - the east-west peninsula in Qiandaohu Lake, it is surrounded by water on the south, west and north side. And its east is connected with the urban road which leads to Qiandaohu Town. This ideal location provides convenient transportation and complete facilities for the hotel. Meanwhile, the hotel itself will improve the leisure environment of Chun'an County. With mountains and lake around, it will be excellent for leisure and vocation.

Planning

To adapt to the surrounding environment, the entrance, road, layout of the buildings as well as the landscape for the project are well considered to make the buildings "grow" on the site naturally.

Entrance: The site connects with the urban road on the east side where the land is flat. The entrance for the resort is set here with a tennis court and a small

gym center near around. The buildings are partly sunken to hide in the green shade. In this way, it has made fully use of the land and provided open and favorable view for the entrance area.

Road: The roads inside the site follow the topography, and the main road is planned along the high land to well connect with the sub roads as well as the buildings. The road system is clear. Walkway is set along the lake to enjoy a great lake view and connect several landscape nodes together. Natural landscape thus connects with the man-made landscapes perfectly to provide a three-dimensional eco system and an ideal place for walk.

Space Layout: The hotel buildings are well arranged to separate from the holiday villas. The hotel is in the center of the site due to the flat land and spacious space here. Main entrance is also set here to give a grand impression and reduce the move of the earth. Its location also ensures both openness and privacy. And according to road planning, holiday villas are arranged in the west to be relatively independent, free of disturbance from other vehicles and people which will be good for safety and management. In addition, since the site is surrounded by water on three sides, it enables all the villas to have a great lake view.

Architectural Design

Skyline Control: The buildings are arranged to follow the topography. In addition, the heights of the buildings are controlled. The central part of the hotel has five floors, and the rest parts are four-storey high. And the villas are designed with 2-3 floors. In this way, buildings can be combined with the mountains perfectly. They stand high and low according to the topography in setback style, present a harmonious and beautiful skyline when seen from the lake.

Single Building Design: The design is based on two principles: try to reduce the damage to the natural landscape of Thousand Island Lake and make the buildings part of the new landscape system; make full use of the natural landscape, not only turning it the beautiful settings but also the source for pleasant experience.

House Layout

For the single house design, it highlights three themes - landscape, holiday and orientation. The hallway, rooms, kitchen, bathroom as well as the courtyard and private garden are arranged according to the lake and mountain views. Sitting room, dining room, breakfast room and kitchen are well connected, courtyard space is shared by the family and friends, every bedroom is designed with a bathroom, windows are set for around views, and sunshine terrace enables one to enjoy beautiful views.

Landscape Design

The designers considered the nature of the existing landscape, setting swimming pools, landscape architectures and waterfront platform according to the topography. Natural landscape thus is combined with the man-made landscape to meet the requirements for leisure and entertainment. Existing plant species are retained and rearranged to form diversified ecological communities together with other colorful and seasonal plants.

PARADISE FOR LEISURE, HEALTH CARE AND TOURISM | Tangshan Xiangyun Island

得天独厚的休闲养生旅游项目—— 唐山湾祥云岛

项目地点：中国河北省唐山市
业主：北京国泰上城置业有限公司
规划设计：凯佳李建筑设计事务所
北京山海树规划建筑设计有限公司
设计人员：克雷•沃格尔，AIA
李春光
总用地面积：770 558 m^2
总建筑面积：1 155 836 m^2
建筑密度：20%
容 积 率：1.5
绿 化 率：30%

Location: Tangshan, China
Client: Beijing Guotai Shangcheng Property Co., Ltd.
Planning and Design: KaziaLi Design Collaborative
Beijing Shanhaishu Planning & Architectural Design Co., Ltd.
Designer: Clay Vogel, AIA
Li Chunguang
Total Land Area: 770,558 m^2
Total Floor Area: 1,155,836 m^2
Building Density: 20%
Plot Ratio: 1.5
Greening Ratio: 30%

项目概况

唐山湾旅游区距北京两小时车程，距离天津1.5小时车程，距离唐山0.5小时车程，距离秦皇岛1小时车程，其规划意在与北戴河旅游区连接形成秦皇岛一乐亭海滨度假带，形成东北亚海路休闲圈、环渤海休闲圈及京津休闲圈三个层面。

祥云岛旅游区位于唐山湾国际旅游岛渤海湾黄金海岸处，平面形状呈保龄球状，其与附近的金银滩、菩提岛、月坨岛、金沙岛等大小岛屿共同形成了中国罕见的最独特的双岸并延地貌格局。

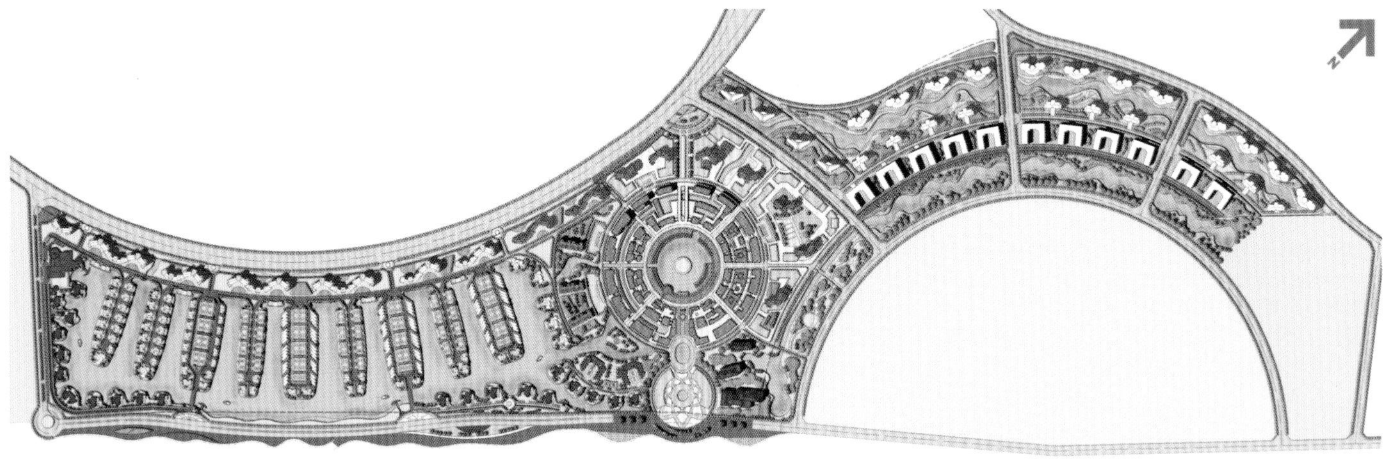

总平面图 Site Plan

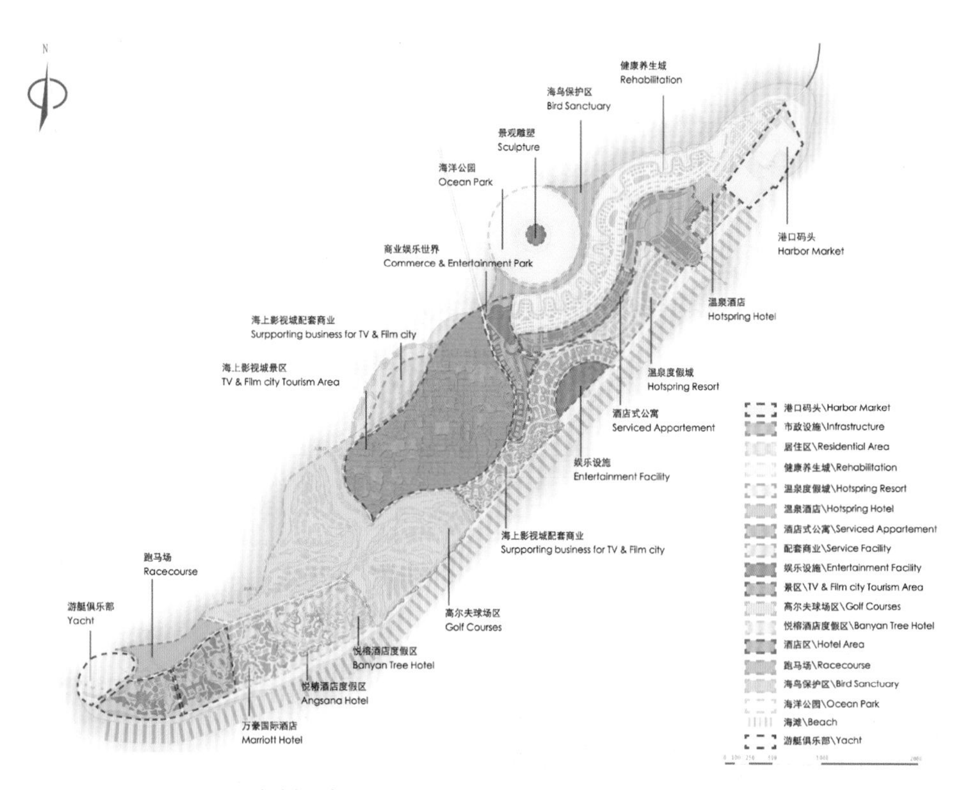

规划设计图 Conceptual Planning Drawing

规划布局

受唐山九大国际品牌委托，总工程项目将包括一座新的跨海大桥以及一个独特的综合性的文化旅游胜地。旅游区由海上影视城、高尔夫主题公园和高端酒店区三大部分组成，其规划设计旨在达到可持续生态发展的目标。海上影视城以横店圆明园为蓝本，维持建筑与水之间现有的空间关系，达到永恒的平衡；位于祥云岛的两个锦标赛高尔夫球场借鉴古老的苏格兰球场而建；岛屿西端的高端酒店区占地面积约20 234 282 m^2，设有五座五星酒店，其中就有排名世界前十的奢华酒店品牌——新加坡悦榕庄，该品牌现已拥有20家顶级度假酒店，其中六家位于泰国普吉岛。

该项目正位于祥云岛高尔夫球区东侧，海上影视城以南，其包括两大部分，一期工程是销售中心，目的是服务于销售，主要功能是接待来岛上看房的购买者，为其提供全方位高档次的体验服务。其次服务于海滨浴场，为旅游度假人群提供高档次的温泉娱乐服务，同时保证冬季同样可以使用。二期主要

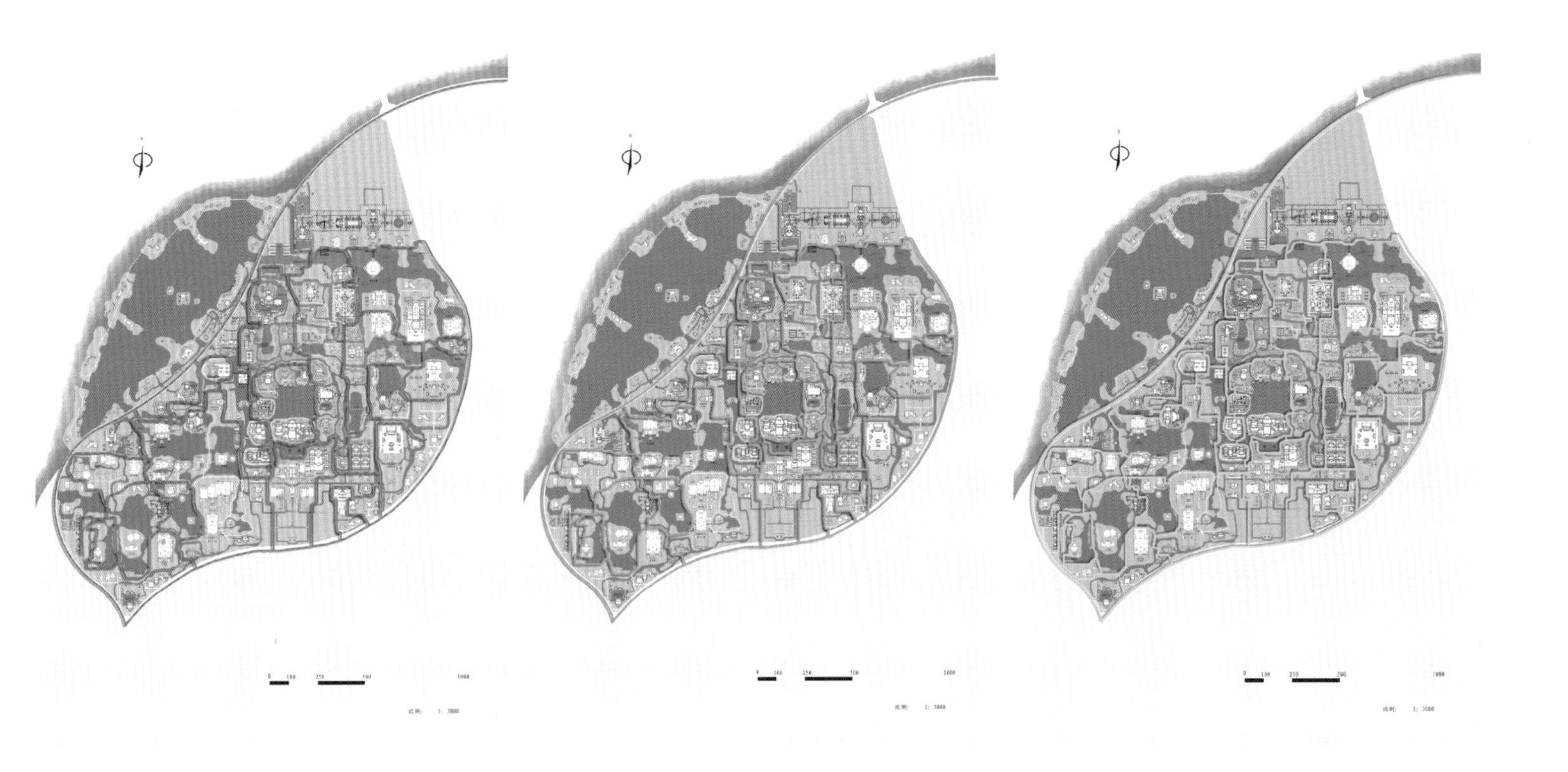

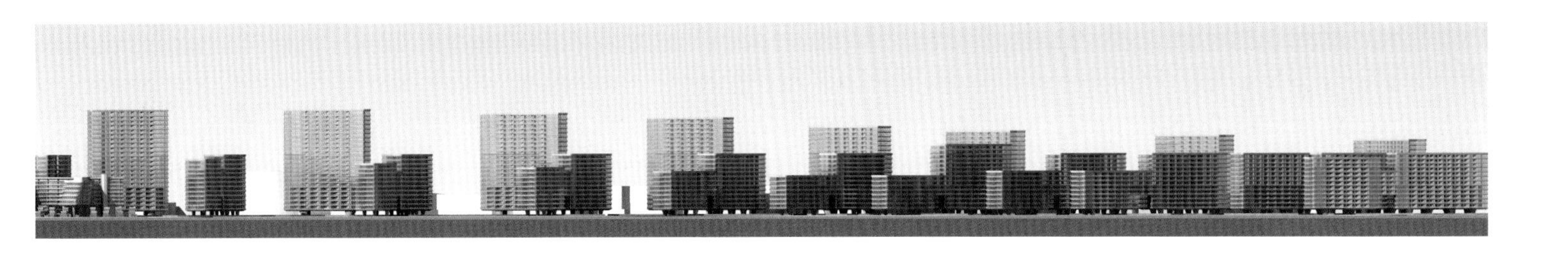

包括点式公寓，独栋、联排别墅，酒吧商业街等。

分散式商业布局和集中式商业布局各有利弊，项目采用商业集中布置，是因为集中布置避免过多分散的商业产生同质竞争，同时形成各种商业互相借势的局面。因此大型集中式的商业却更适合滨海旅游大量集中人群的情况。

中心商业区集中规划选址选择与主要来岛交通靠近，远离高尔夫球区，减少互相影响；有一定临海资源较佳。其根据“太极”布局概念，使建筑呈现出中心向四周扩散的建筑布局，很好地分割出路网形态，方便车流、人流的进出。此外，商业街还分成以水街为主、以小街小巷为主、部分集中休闲区集合自然景观与绿地、商业围合广场形体提供更多的活动空间等商业形态。

建筑设计

水路交通：鉴于海岛地势低洼，岛外的土方又比较稀缺，岛内充分利用挖沙造水景的方式来进行自身的土方平衡，这样大大减低了土方运输所造成的碳排放，而且也是一种良好的可持续性的土方平衡方案，由此便形成了岛内特有的水路交通。

停车位标高的确定：现场的场地标高约在1 m左右，根据海岛的条件，主路及建筑的首层标高在4 m左右，这样自然形成了一个开敞的地下夹层用于停车。

可持续发展设计：用传统的混凝土、传统的沥青、可渗透的铺路材料；用景观植被去减缓；过滤器净化，并且过滤流动的雨水；设置屋顶自然的条纹等。

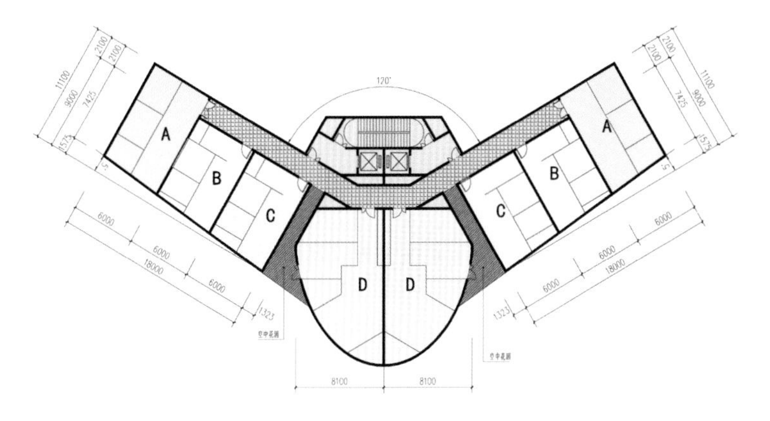

标准层平面图 Plan for Standard Floor

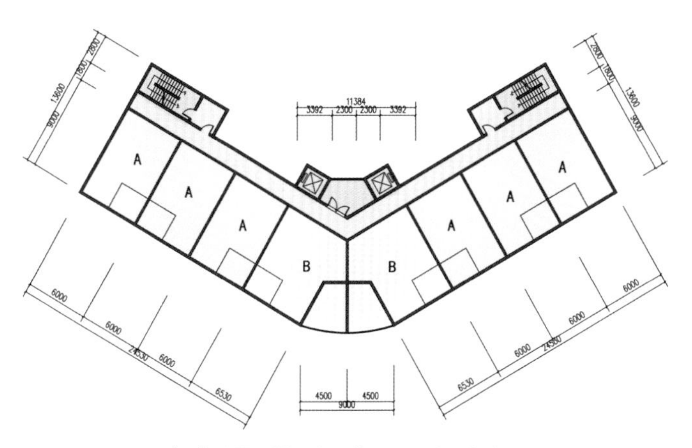

标准层平面图 Plan for Standard Floor

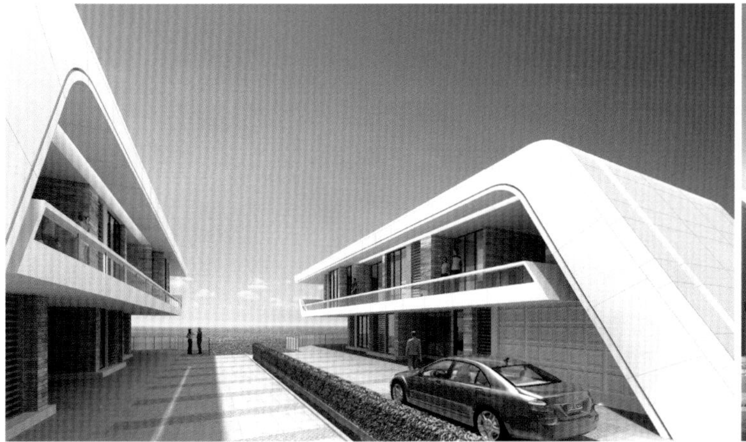

Profile

Tangshan Bay Tourism area is located on the northeastern seaside, taking 2 hours-drive to Beijing, 1.5 hours to Tianjin, 0.5 hour to Tangshan and 1 hour to Qinhuangdao. It is planned to form a seaside tourist destination from Qinhuangdao to Leting together with Beidaihe Tourism area, which includes three sections: Northeast Yahailu recreation rim, Bohai recreation rim and Beijing-Tianjin recreation rim. Xiangyun Island Tourism area is located on Tangshan Bay International Tourism Island, the the golden coast of Bohai Bay. The site looks like a bowling ball which forms the unique "double coast" together with Jinyin Beach, Bodhi Island, Yuetuo Island, Golden Sand Island, etc.

Planning

As commissioned by the Tangshan Nine-International Brand, this project includes a new sea-crossing bridge, as well as the community planning and design of a cohesive, unique, and

culturally-branded tourist destination - which includes three distinct sections: Sea Studios, Golf Theme Park, and a high-end hotel district - while achieving sustainable eco-development goals. The Sea Studios is based on palinspastic map of the Old Summer Palace from Hengdian Group, and as such, maintains the original spatial relationships between the buildings and water, creating a pleasing balance with a sense of timelessness. The two championship golf courses on the Island are based on ancient golf courses in Scotland. The west end of Xiangyun Island is slated for a high-end hotel community with five 5-star hotels, comprising about 20,234,282 m^2 of land. Included in the hotel community is one of the world's top ten luxury brands: Banyan Tree of Singapore, which has over 20 top Holiday hotels, six in Phuket, Thailand.

The project is just on the east side of Xiangyun Island Golf Course and the south of Sea Studios. The project consists of two parts. Phase I is the sales center for reception and providing a high-end experience. It also serves the tourists of the seaside bathing area with hot springs all year round. Phase II includes point-type apartments, single-family villas, townhouses, bar street, etc.

Distributed commercial arrangements and centralized commercial arrangements have their own pros and cons. This project applies the latter to avoid distributed competition and establish mutual benefits. It will help to attract more tourists to the area.

The commercial center is set near to the entrance and far away from the golf court to avoid mutual disturbance. It can enjoy the seaside resources. Buildings are arranged in a "Tai Ji" layout with a clear traffic system. In addition, the commercial area not only includes the canals, side streets and alleys but also consists of commercial plazas which create leisurely spaces for visitors.

Architectural Design

Waterway Traffic: Due to the low-lying land and the lack of earth, the sand beaches have been excavated to create numerous waterscapes and provide surplus land which can be used to build the project. In this way, it not only reduces carbon emission during earth transportation, it also creates a balance between the earthwork and forms the unique waterways inside the island.

The Standard Height of the Parking Lot: the elevation of the site is about 1m. According to the island conditions, the main road and the first floor of the building are 4m high to provide an open space for car parking.

Sustainable Development Design: Traditional concrete, pitch, and penetrable materials are used for ground cover. Green landscapes are used to soften the hard feelings. In addition, filters are applied to recycle rainwater. Roofs are designed with natural strips.

NINE SPACES, PERFECT INTERPRETATION OF ZEN CULTURE TOURISM

| Constructional Detailed Plan for Zen Exposition Park, Yichun

九度空间完美诠释禅文化旅游—— 宜春市禅都文化博览园修建性规划项目

项目地点：中国江西省宜春市
开 发 商：宜春城市投资集团
规划设计：麟德旅游规划顾问有限公司
规划人员：许文龙、张立志、王屹、黄赛婵、刘娇丽、郁培媛
总占地面积：2 100 000 m^2
总建筑面积：432 000 m^2
景区部分建筑面积：120 000 m^2
地产区建筑面积：312 000 m^2
地产区容积率：2.3
地产区绿化率：35%

Location: Yichun, Jiangxi, China
Developer: Yichun Urban Construction and Investment Group
Planning and Design: Luntak Tourism Planning Consultant Co., Ltd.
Planner: Xu Wenlong, Zhang Lizhi, Wang Yi, Huang Saichan, Liu Jiaoli, Yu Peiyuan
Total Land Area: 2,100,000 m^2
Total Floor Area: 432,000 m^2
Floor Area (Landscape Area): 120,000 m^2
Floor Area (Estate Area): 312,000 m^2
Plot Ratio (Estate Area): 2.3
Greening Ratio (Estate Area): 35%

项目概况

宜春市禅都文化博览园位于江西省宜春市，该项目占地面积约2 100 000 m^2。该规划通过评审后即完成立项，目前已进入施工阶段。

规划布局

规划理念：宜春是“马祖兴丛林、百丈立清规”的起源处，又是禅宗“一花五叶”中的三宗发祥地，故称为“禅宗圣地”。本方案充分考虑佛的内涵、禅的主题及宜春的城市特色，把禅宗的“一花五叶”与宜春起源的“丛林清规”自然融合。

规划主题：以禅文化为主线，贯穿整个景区。

总体布局：打造九度空间，即“一花、五叶、三丛林”。分别以禅心、禅念、禅地、禅祖、禅寺、禅业、禅客、禅居、禅乡为依托，以悟禅、入禅、识禅、观禅、修禅、施禅、客禅、城禅、乡禅为内容，形成具有不同禅意境的九个功能空间。九个功能空间整体形成旅游和居住两大片区：一花五叶形成旅游片区，一花辐射五叶，五叶环绕一花，犹如佛光法轮之形，暗合禅宗圆融之义，阐释宜春明月之心，全面展示佛心禅、世俗禅和宜春禅。三大丛林形成居住片区，因禅文化旅游而生，环绕“一花五叶”分布，包含主题酒店区、高尚居住区、乡土民俗区。林在楼间，楼在林中，演绎世俗之禅，形成三大特色人居环境。

规则总平面图 Regulation Plan

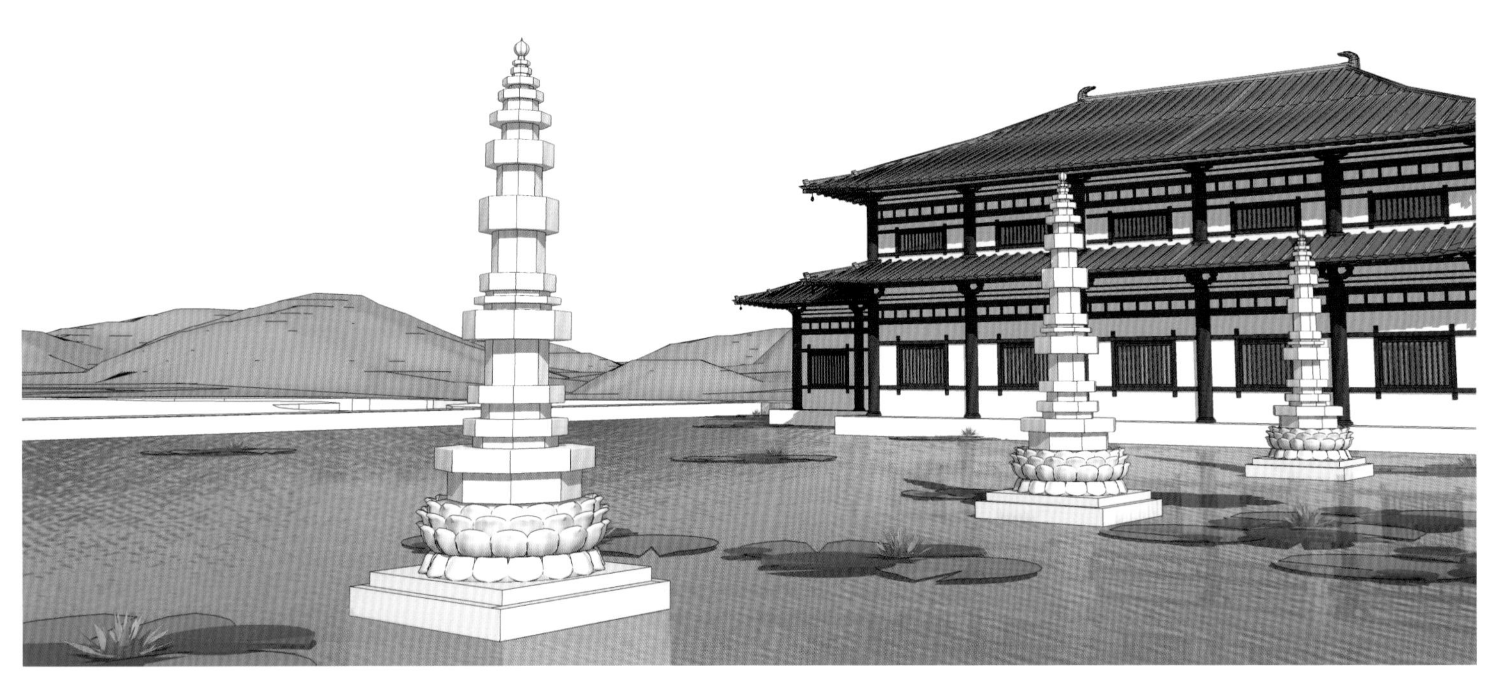

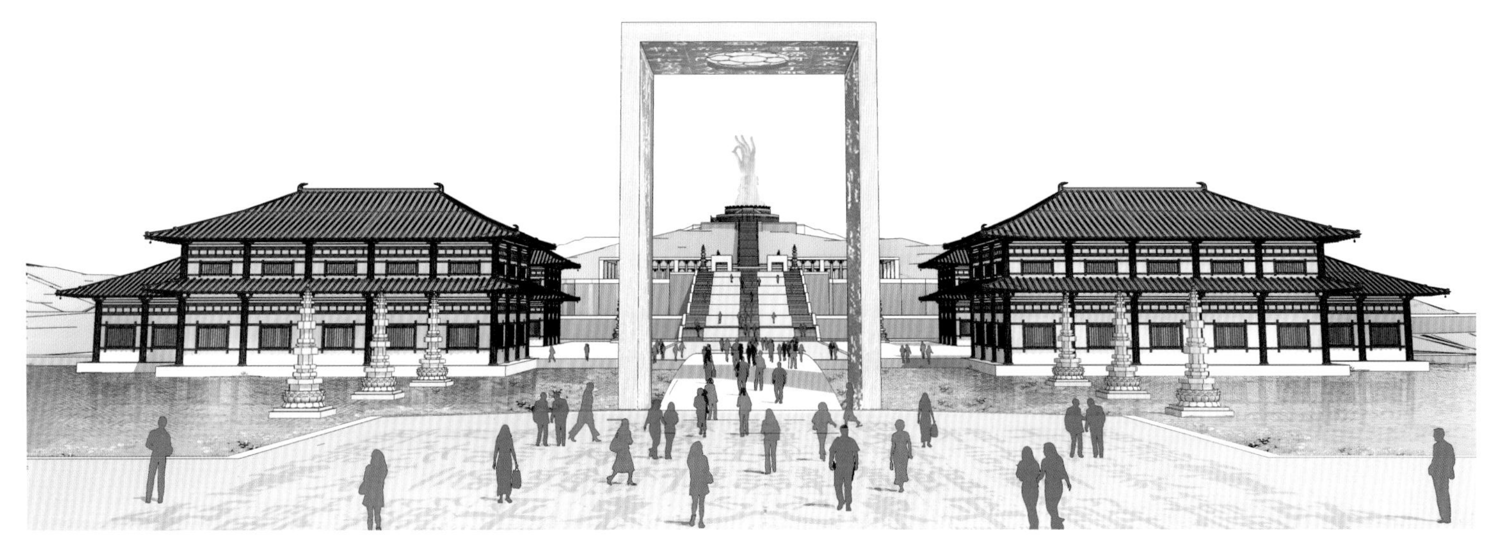

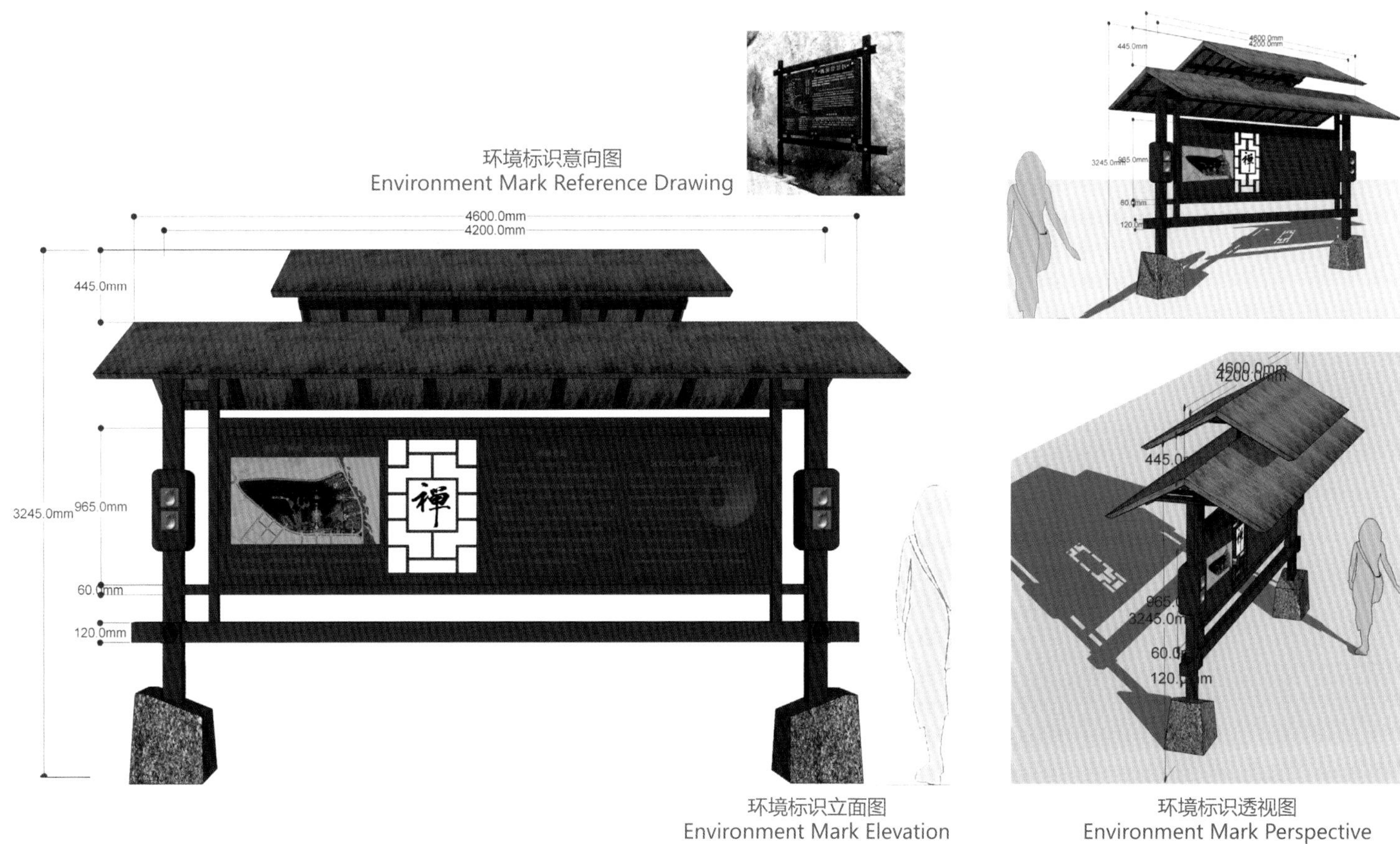

环境标识意向图
Environment Mark Reference Drawing

环境标识立面图
Environment Mark Elevation

环境标识透视图
Environment Mark Perspective

设计理念

主题酒店区的设计理念：以禅客为依托，以客禅为内容，突出一个“客”字，客心、客缘、客禅月。以禅月酒店为核心，把宜春的禅及月亮文化结合。开发以禅月为主题的度假酒店及其配套设施，营造“把酒问月，满目禅风”空间意境。

高尚居住区的设计理念：以禅居为依托，以城禅为内容，突出一个“城”字，城里、城外、城中禅。以禅意社区为核心，引入禅的文化，通过建筑、景观及园林的设计与营造，配套完善的社区生活及商业设施，营造“依山望水，禅境人家”空间意境。

乡土民俗区的设计理念：以禅乡为依托，以乡禅为内容，突出一个“乡”字，乡村、乡俗、乡间禅。以禅园新村为核心，把乡村民俗与禅文化结合，建设富有禅意的特色乡村。营造“游村览俗，心悟禅机”空间意境。"

Profile

Zen Exposition Park is located in Yichun City of Jiangxi Province with a land area of 2,100,000 m^2. After passing the evaluation and getting the approval, it is now under construction.

Planning

Planning idea: Yichun is the origin for the story – "Mazu establishes temples, and Baizhang sets the regulations". And it is also the birthplace for three branches of the Zen Buddhism. The plan has fully considered the theme and culture of Zen as well as the characteristics of the city to combine "temples and Buddhist regulations" with Zen Buddhism.

Theme: Zen culture penetrates in every corner of the park.

Overall layout: it tries to create nine spaces which refer to "one flower (Bodhidharma), five leaves (branches of Zen Buddhism) and three temples".

Nine different functional spaces are designed according to different themes and contents. These spaces are divided into two parts for tourism and living: "one flower and five leaves" form the tourism area, looking like the Dharma Chakra which implies the harmony in Zen Buddhism. It showcases the Zen culture of Yichun city. "Three temples" form the living area which composes of theme hotels, high-end neighborhoods and village area. Buildings and temples stand between each other to combine the spirit of Zen into daily life.

Design Idea

Theme hotels: relying on Zen culture, it emphasizes the word "guest". Zen culture and moon culture of Yichun are combined in the theme hotels to present unforgettable experience.

High-end neighborhood: based on the living culture of Zen style, it highlights the word "city". Zen culture is introduced in the neighborhood and penetrates into the design of buildings, landscapes, gardens and the supporting facilities.

Village area: based on the reputation of "Zen village" and with the content of Zen culture, it highlights the word "village". Combining folk custom with Zen culture, it has established a typical new village in this Zen Park. Tourists can experience the strong Zen ambiance here.

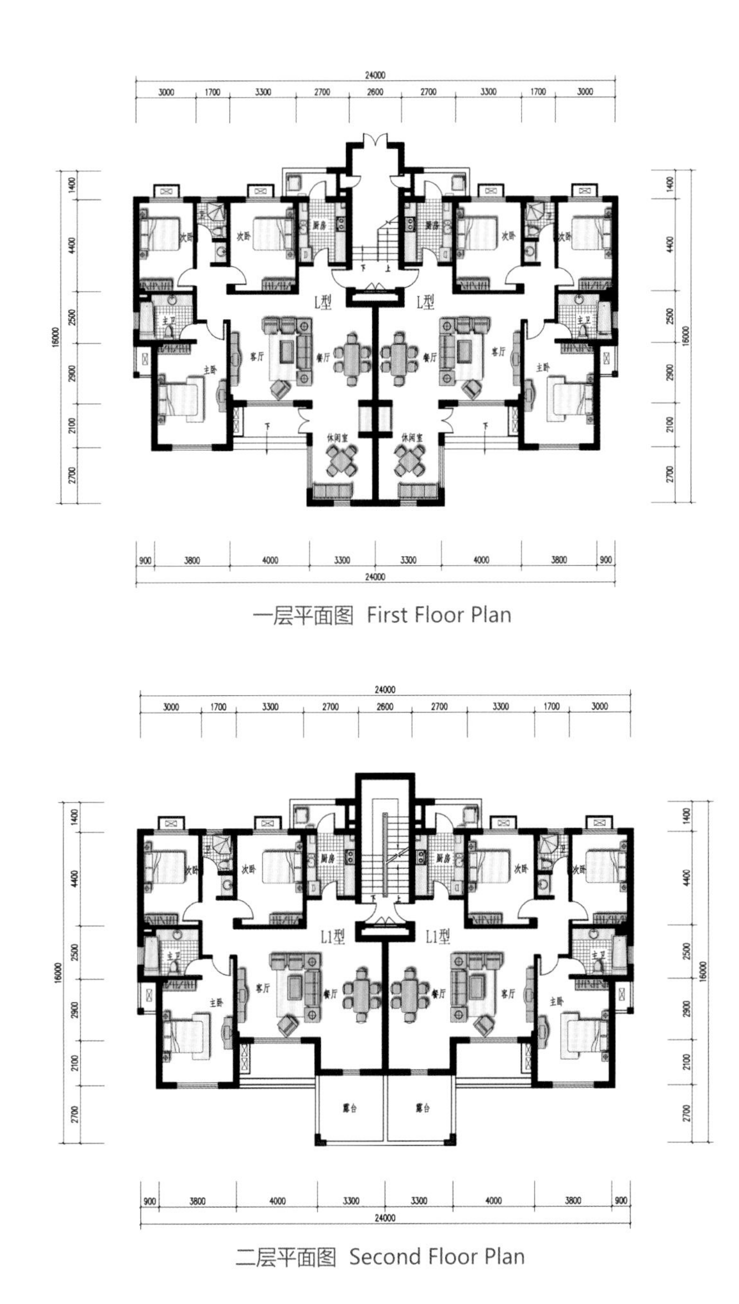

一层平面图 First Floor Plan

二层平面图 Second Floor Plan

城市旅游——城市性格决定旅游特色

麟德旅游规划顾问有限公司董事长 杨力民

■ 作者简介

杨力民教授，著名学者、旅游策划专家，长期从事旅游、美学、历史学等多学科研究。杨教授1945年生于陕西西安，1965年毕业于西北轻工业学院并留校任教，1978—1993年任陕西省旅游局国际市场开发处副处长、处长，1993年初定居香港，创办麟德企业公司，现任麟德企业董事长，同时担任河南省政协常委、香港省级政协委员联谊会常务理事、深圳市总商会（工商联）荣誉副会长、河南海外联谊会副会长、黑龙江海外联谊会副会长、中国光彩事业促进会理事等多项职位。

作为著名旅游策划专家，杨力民教授同时受聘担任五十余个省市的政府旅游高级顾问，曾为省市相关部门作"创意旅游"、"区域旅游的开发设计"、"旅游市场营销"等旅游学术报告百余场，对当地如何以旅游为支点推动区域经济的发展等问题进行剖析阐述，一致受到极高的评价。杨教授还多次受邀参加高级别的国际旅游论坛并发表演讲，著有《中国旅游大辞典》、《中国古代瓦当艺术》、《创意旅游》、《城市营销》、《国学旅游》等专业畅销书籍，并亲自主导过上海世博会中国馆中标单位、中国第一届旅游交易会总体设计等项目。

自从中国第一个王朝夏朝建立，定都阳城，"城市"这个词汇进入帝国的视野以来就带着如负千钧的震撼和责任。"城市"最重要的两种功能：一是满足安居自守的使命感，二是满足对物质经济的渴望，"城市"不可避免地被这两样功能驱策前行。"城市"是一个复杂而抽象的概念，过去时代沉淀下来的社会文化、民俗世态、风物人情等都潜伏在城市日常生活的底层，细微而持续地融合。一言以蔽之，"城市"作为中国文化发展的前后相续的见证，记录了中华五千年史——从战国以前的奴隶制国家形态发展到今天大一统国家的建立，一个完整的城市阶层历经数千年孕育最终成型。总之，绕开"城市"就无法理解帝国、无法理解帝国精神。

在新的时代背景下，城市化进程如此之快，城市的功能早已经远远超出原始的定义，而是涵盖了从居住、教育、商业、金融、娱乐等各个领域，其中旅游度假作为近些年新兴的城市功能越来越受到人们的注意。例如三亚在短短的几年时间迅速发展成为全国的焦点，旅游对三亚整体提升作用功不可没。可以想象，未来社会随着城乡一体化的迅速发展，城市经济时代的全面到来并不遥远，城市必将成为旅游市场竞争的主体。竞争需要的是特有的冷静、克制和精确，这打破了学术与通俗的分界，旅游作为城市发展的新兴引擎，通过对城市的社会风貌深刻、新颖的剖析和解释，做出见微知著、融会贯通的整合，有着指导城市发展的巨大作用。如今新时代旅游正要融入城市，要为城市经济大局服务，要为城市经济作贡献，要做城市大旅游，其将旅游作为城市发展的重要动力，而城市则作为旅游发展的依托。

城市可以拟人化，可以像人一样具备性格。怎样定位属于自己的城市性格，不仅仅是有关公众对城市的认同，更是指导城市社会经济文化发展的重要依据。当城市策划者没有足够的准备把握城市的性格，树立城市肌理和公众的归属感时，策划城市就存在相当的难度。城市意象以最难以确定的存在和最具差异化的感知程度存在于每个市民的心中，策划城市就要深知所在城市的历史、现在和未来的定位，要了解公众对城市感知上的不可莫测的心理。城市都渴望通过对自身性格的精华提炼来定位形象，如果一个城市的定位没有展现她的性格，对自我挖掘不够充分，对区别于其他城市的特点也就认知不足，情商的得分也不会高。

城市的形象定位中缺乏个性是降低城市情商的一个错误，随着城市化进程的快速推进，很多城市盲目模仿、攀比，这种盲目山寨的做法只会谋杀城市性格，其结果就是城市在自我发展定位时失去个性，在越来越多的城市呈现出高度近似气息的情况下，城市发展要开辟一块新天地就必须有超然独到的眼光，这就是要抓住每一座城市自己的性格。旅游对城市的研究，特别是对城市性格的把握，是至关重要的；从城市的性格角度看旅游，而不是从技术层面的角度出发，这一点异于以观光开发为中心的价值观。而重新归纳、综合、试图从城市发展的角度看旅游策划的研究风格也不同于一般主流的眼光。

从来没有人会把自己单纯地束缚在一张乏味的旅游地图上，人们怀着对城市性格的理解，在不断的尝试中寻求转化的灵感。精明，温柔，大气，坚韧，这些都可以归结为一个城市的性格，城市性格不单指城市里居民情感外露带给外来游客的第一印象，也特指一个城市的每个生活细节、人文特点，这都可以从一个城市的风土人情中一窥究竟。北方城市性格大多男性化，南方城市则多数偏女性化。杭州的性格来自西湖，南京的性格来自紫金山，厦门的性格来自鼓浪屿，青岛的性格来自崂山。城市性格不仅能指导城市策划的核心思想，更可以指导具体的旅游产品。上海人性格精明，策划上海的旅游就要从上海人的角度出发，计算到每个细微的角落；北京人讲究大爷气质，北京的旅游产品要注重气场能表现出气势磅礴；西安人性格坚韧，在西安旅行就要给人一种厚重感……中国人相信，性格能反应人的灵性和觉悟。这一点就仿佛城市性格对旅游的塑造，两者间有一种不可言喻的共通。

游客是带着休闲观光的心情来到城市的，如何彻底放松是他们最关心的问题，如何享受与众不公的，与其他城市差异化的旅游产品是他们最关心的问题。城市性格最大的作用就是要叫游客在经历过旅行之后能够快速地建立起城市形象，当城市作为一种新兴旅游产品出现在人们的面前时，其品牌的忠诚度其实与其他产品是并无区别的。人们在经历过一次美好的城市旅游之后，除了自己会在未来的日子里重新品位，还会有重复体验的强烈愿望，并且不由自主地将自己的美好经历告诉身边的人，这样就等于向其他人推荐了城市的旅游品牌形象。虽然城市这种旅游产品不像柴米油盐酱醋茶那样是生活必需品，但是今天只要把握城市性格那么城市被重复旅游和多次消费的可行性就越来越大。

透过看似平静的海面探寻城市传统文化的大陆架如何突降为海床，从而辨析城市旅游策划如何起于青萍之末，把休闲的旅游和博通的城市性格熔于一炉，成为大众的精神美食可能是我们做城市旅游以后要重点考虑的问题。

Urban Tourism -The Theme is Decided By City Style

Yang Limin, President of Shenzhen Luntak Tourism Planning & Consultant Co., Ltd.

Profile:

Professor Yang Limin, eminent scholar, expert in tourism planning, has long been committed to the researches on tourism, aesthetics, and history. Born in 1945 in Xi'an, Shaanxi Province, professor Yang Limin graduated from Northwest Institute of Light Industry and worked there as a teacher in 1965. During 1978 to 1993, he was the deputy director and director of the International Development Department of Shaanxi Tourist Administration. And in early 1993, he settled down in Hong Kong and established Luntak Enterprise. He is now the president of Luntak Enterprise, and is assigned to be member of the Standing Committee of the Political Bureau of Henan Province, executive director of Hong Kong Provincial Association of CPPCC Members, honorary vice-chairman of Shenzhen General Chamber of Commerce, vice-chairman of Henan Overseas Friendship Association, vice-chairman of Heilongjiang Overseas Friendship Association, director of China Society for Promotion of the Guangcai Programme, etc.

As a renowned expert in tourism planning, Yang is the senior consultants for the government and tourism institutes for about fifty cities. More than one hundred lectures on tourism were made by him, such as "Innovative Tourism", "Development and Design of Tourism Region", "Tourism Marketing" and so on, helping to promote the tourism development and drive the economy in these regions. All of these lectures are highly praised. Yang has also been invited to join in some high-class international tourism forums and made speeches. His important publications include The Chinese Travel Dictionary, The Art of Eaves Tile in Ancient China, Innovative Tourism, City Marketing, Chinese Cultural Tour, etc. And his representative projects are China Pavilion of Shanghai Expo 2010 and master plan for the 1st China's Travel Market.

Since the establishment of the first Chinese Empire - Xia Dynasty, with its capital in Yangcheng City, the word "city" had entered into the history and began to bear great responsibilities. Two basic objectives of a city are for settlement and self-protection, and for pursuing of material economy. The development of cities is just driven by these two objectives. "City" is an abstract and complicated concept, with culture, scenery, local customs and mores behind urban daily life. All of these elements are combined perfectly and developed together. In short, as the witness of the development for Chinese culture, "city" has recorded China's 5,000 years' long history. A complete city system is formed till now. The understanding of an empire and its spirit cannot go without the word "city".

Nowadays, with high-speed urbanization, the functions of cities have been far beyond the original. More are included in "city", such as living, education, commerce, finance, entertainment and so on. As the new function, tourism has drawn more and more attentions. For example, tourism made a great difference in Sanya within several years, changing the city the focus of China. As you can imagine, with the rapid urban-rural integration, the times of urban economy is approaching, and cities will become hotspots of the travel market. Competition needs special calm, control and accuracy which push the boundaries between academic and popular. Tourism thus serves as the new engine that drives the development of the cities, reanalyzes and reinterprets the urban styles and features, and integrates multi elements together. Now, modern tourism is ready to integrate in cities and serve for the economic development. City will be the carrier of tourism, and in return, tourism will drive the development of cities.

Cities can be personified with typical styles. Well-defined city style will not only promote the city image to the public but also guide the development of social economy and culture. When the city style cannot be defined to form the urban context and provide the feeling of belongingness, urban planning is hard to achieve. Since different people hold different impressions to a city, the urban planning must be based on the deep research of the history, present and the future orientation of the city. The orientation of a city should reflect its own style, showcasing its characteristics and differences.

Lack of personality will greatly decrease the EQ of a city. With the rapid process of urbanization, many cities just imitate and compare with other cities blindly which has killed their own personalities and lost the orientations. Thus to present a city of brand new, the urban planning must be unique with foresight to keep and enhance the individuality of the city. Tourism plays an important role in this process. Defining the theme of tourism according to the personality of the city is quite different from other tourism development. It will put tourism planning into the whole urban planning to recollect information and find typical style of a city.

No one will tie himself just to a travel map. People will try to find the inspiration of change with their understandings to a city. Shrewd, warmness, elegance or toughness, all these words can be used to express the personality of a city. The personality not means the first impression the city gives to the citizens but also appears in the life details, cultural characteristics and local customs. The cities in North China seem masculine, while the southern cities are usually feminine. Hangzhou City gets its personality from the West Lake, Nanjing from Purple Mountain, Xiamen from Gulangyu Island, and Qingdao from Laoshan Mountain. The personality of a city will not only guide the urban planning but also decide the theme of the tourism. Shanghainese are shrewd and the planning of Shanghai tourism should pay attention to every detail and corner; Peking people highlight greatness, thus the tourism products should be magnificent; Xi'an people are tough and tensile, so Xi'an tour should give the sense of steadiness. Chinese people believe that personality reflects the spirit and thought. It is just like the relationship between city style and tourism planning.

Tourists come to a city for sight-seeing and relaxation. How to relax themselves to the most is what they concern most. And the different tourism experience is what they want when going to a new city. The personality of a city helps the tourists to establish their expressions to the city. When one enjoys a pleasant journey, he will remind of it in the following days and may hope to experience again. He will also share this experience with people around and recommend it to others. Even though the tourism products are not so essential like other necessities, city with typical style will win more visitors.

How to find the traditional culture from the surface of a city and combine urban tourism with city style, making it the spiritual food us will be the things we should consider first when doing tourism planning.

INTELLIGENT INTERFLOW RELAXATION SPACE | Ningbo Yinzhou Talent Apartment

智慧 交流 休闲 空间——宁波鄞州区人才公寓

项目地点：中国浙江省宁波市鄞州区
开 发 商：宁波鄞州区新城房地产开发有限公司
建筑设计：DC国际
宁波市鄞州建筑设计院
总用地面积：33 375 m^2
总建筑面积：107 800 m^2
容 积 率：2.50
绿 地 率：30%

Location: Yinzhou, Ningbo, Zhejiang, China
Developer: Ningbo Yinzhou Xincheng Real Estate Development Co., Ltd.
Architectural Design: DC Alliance
Ningbo Yinzhou Architecture Design Institute
Total Land Area: 33,375 m^2
Total Floor Area: 107,800 m^2
Plot Ratio: 2.50
Greening Ratio: 30%

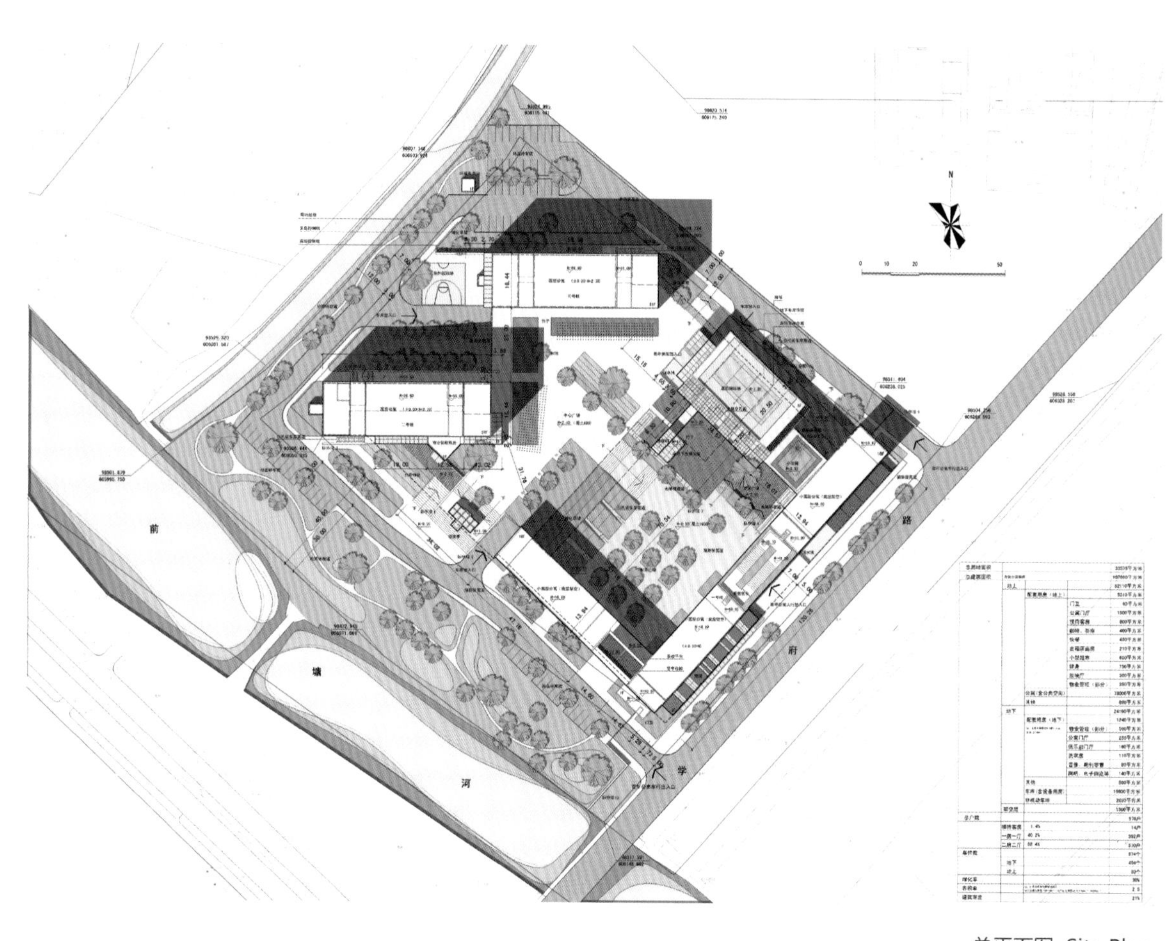

总平面图 Site Plan

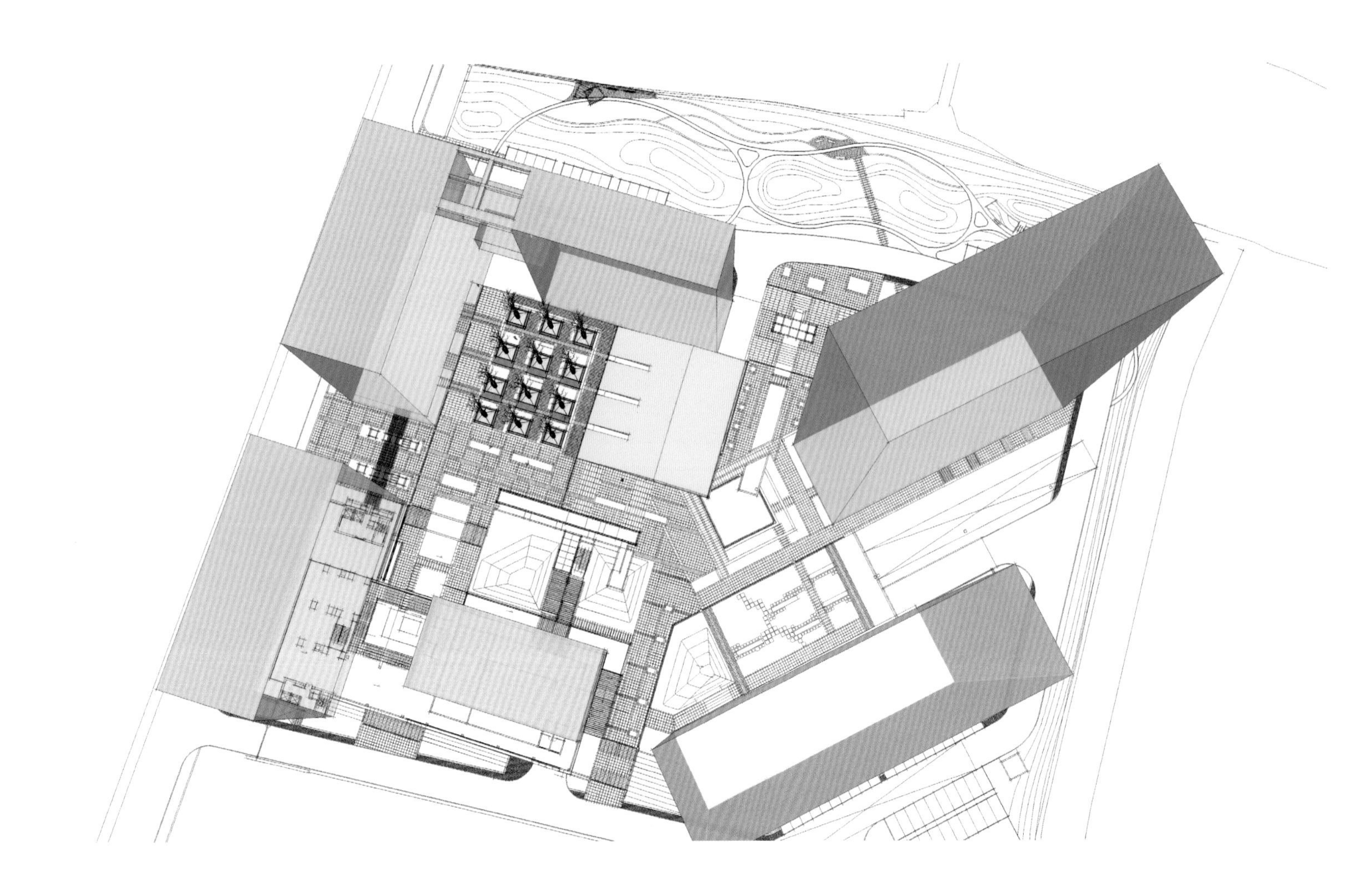

项目概况

基地位于宁波市鄞州区钟公庙街道，东为学府路，南为前塘河，北面为教师公寓，泰康东路与学府路交叉口的北侧，地处鄞州新城区高教园区边缘，整体氛围相当契合项目的定位，并提供了相应层次的文化背景、辐射人群以及设计工作的分析基础，区位环境优越。项目总建筑面积107 800 m^2，其中地上82 110 m^2，地下24 190 m^2。

规划布局

本项目希望作为宁波住宅市场的一种新型高端产品在鄞州区率先试行，因此其准确定位为“人才+公寓”，为特定受众人群提供特定的生活方式，并为这种生活方式树立建筑模板，规划为各企业引进的人才提供约1000户小型单身公寓。整个项目的地上建筑包括单身公寓、商业用房、青年俱乐部、物管用房等；地下为机动车停车库、非机动车停车库、设备用房等。

建筑设计

整个建筑的外立面简洁、时尚，富有现代感，大面积玻璃的运用是本项目的一大特色，而富有规则的外立面结构让整个项目富有几何美感；建筑之间的合理朝向以及个建筑组团之间的合理布局，使得建筑具有良好的视觉和空间感。

户型设计

整个建筑空间布局小巧紧凑，富于情趣；人性化室内动线设计，空间的实用率高；并设置有敞开式的厨房，整体光线通透，空间层次感丰富，景观尽享的网络城市生活提前在此上演；阳台采用多维的视野，丰富感官的享受，景色风光处处纵览无遗。

景观设计

项目基地内高密度的栽植各类树种，落叶与长青树种交错，营造各个季节不同特色的景致，并通过树木不同的特性进行种植，形成围合的林荫道。同时还有小型开放空间和小树林，充分注重建筑物与树木的协调相互掩映，形成景观氛围与空间利用的双重效果。

Profile

Located at Zhonggongmiao Street, Yinzhou, Ningbo, the site is adjacent to Xuefu Road on the east, Qiantang River on the south, teacher's apartment on the north and is north to the intersection of Taikang East Road and Xuefu Road. Situated at the edge of Yinzhou Higher Education Park, the overall atmosphere quite fit the positioning of the project, and provides the appropriate level of cultural background, radiation crowd as well as the analysis basis of design work. Total floor area is composed of 82,110 m^2 above the ground and 24,190 m^2 below, 107,800 m^2 in total.

Planning Layout

As a newly high-end product, it hopes to be the first trail in Yinzhou residential market. Its accurate positioning "talent + apartment" is to provide a specific way of life for a specific groups and establish a construction template for this way of life. 1,000 units were provided by this chance. The ground floor of the entire project includes single apartments, commercial buildings, youth clubs and property management rooms, etc. Spaces below contain vehicle garage, non-vehicle garage and equipment rooms.

Architectural Design

The entire building facade is simple, stylish and modern. The use of a large area of glass is a major feature of this project. Well-regulated facade structure makes the whole project full of geometric beauty. Reasonable orientation between the buildings and the proper layout between the groups

make the construction with good visual and spatial perception.

House Layout Design

Space layout of the entire building is compact and full of fun with humane indoor dynamic line design, high utility rate, open kitchen, transparent lighting, rich layering of space and never-failing landscape. Multi-dimensional perspective was presented in front of the balcony, which offers you a rich experience.

Landscape Design

All kinds of trees were planted in the site in high-density. Deciduous and evergreen trees staggered to create a view of different characteristics of each season, and different characteristics of trees planted to form enclosed boulevard. At the same time there are a small open space and woods that focus on the coordination of buildings and trees to set off each other to form the dual effect of landscape atmosphere and space utilization.

THE MAXIMUM VALUE IMPLEMENTATION OF MULTI-OBJECTIVE SYSTEM DESIGN | Liuzhou Woodland

多目标体系设计的最大价值实现—— 柳州华林君邸居住小区

项目地点：中国广西壮族自治区柳州市
项目策划、空间规划、景观规划、建筑设计：广州市金冕建筑设计有限公司
总用地面积：68 268.6 m^2
总建筑面积：116 057 m^2
容积率：1.7
建筑密度：25.42%
绿化率：50%

Location: Liuzhou, Guangxi Zhuang Autonomous Region, China
Project Planning, Space Planning, Landscape Design, Architectural Design: Guangzhou Kingmade Architecture Design Co., Ltd.
Total Land Area: 68,268.6 m^2
Total Floor Area: 116,057 m^2
Plot Ratio: 1.7
Building Density: 25.42%
Greening Ratio: 50%

起居空间系统

保留原138棵大树——为躲避树木，采用了设计、施工都难度较大的弧形住宅建筑，弧线转角为15度、25度、30度、40度、45度等，尽可能的呵护每一棵大树。**超大楼间景观视距**——普通情况下，多层住宅楼距在20~30 m之间，小高层楼距在30~40 m之间，使得80%以上住户获得60~120 m的景观视距。**一流的通风条件**——低风阻系数的自由曲线设计；全一梯两户设计；空中花园完全中空通风设计。**每户6 m高空中花园**——每一分钟都在拉近人和大自然之间距离。**错层式室内空间**——创造温馨家庭生活情趣，增强私密性。**阳光“普”照**——弧线正南北朝向住宅设计，让北侧的厨房、卫生间同样阳光充足，削弱潮湿环境和减少蚊虫滋生。**喇叭形建筑布置**——自然形成入口广场和商业广场喇叭形的欢迎姿态，而且，延展商业界面的同时有效阻隔东侧城市干道的噪音废气，可谓一石三鸟。

配套服务系统

配套设施规划——规划小区主干道东侧为配套设施集中区域，自然形成动静相对分区，加强了设施与设施之间的良好互动，同时利于小区公共空间向私密空间的良好过渡。**“龙胜梯田”中心庭院**——通过华林居底层架空，将会所风情泳池和网球场及“龙胜梯田”水面连成一片，形成华林君邸浑然一体的中心庭院交流空间。**会所是客厅的延伸**——休闲度假式会所、风情泳池以及多处底层架空休闲空间设计，提供了室内、半室内、半室外和室外丰富的、适合不同人群需求的共享交流空间。**配套设施的多样性**——15 000 m^2商业配套、星级幼儿园、golf果岭、风情游泳池、国际标准网球场、乒乓球、桌球、棋牌等。

规则总平面图 Regulation Plan

道路交通系统

人行交通设计：首先是必要性活动——上下班、购物、办事等出行，满足便捷需要；**其次是选择性活动**——散步、玩乐、闲逛等出行，满足舒适需要；**再次是社交性活动**——谈话、沟通、聚会等活动，满足景观和氛围需要。**车行道路设计**：利用弧形形态，从设计上减缓汽车行驶速度，同时简明联系小区北、东两个出入口，高效服务每一个组团。**阳光车库配置**：项目规模虽仅有6.8万多平方米的用地，总建筑面积11.6万m^2，却为了保护原生乔木不得不设计了多达7个车库，更难能可贵的是做到了几乎全都是自然通风采光的半地下车库。

景观环境系统

超过1500 m林中漫步——华林君邸在仅6.8万多平方米的面积内却拥有超过1500 m、连续不断的林荫漫步风景长廊，在这条精心打造的空间“龙”中漫步，你能感受到空间收放、步移景异，40年高大原生树木伴随着你，犹如一支优美乐曲抑扬顿挫，带领你的心情畅游。**建筑外立面景观**——非对称设计，变化丰富的框架似跳动的音符，加上自由曲线建筑界面设计，随四季和一天的阳光变化而光影变化，获得每天的全新形象。**四维立体绿化**——纵、横二维平面绿化精心设计，加上每家每户空中花园绿化，以及40年高大树木和新树发芽的时间维，构筑丰富多样的立体绿化景观。**无水不成居**——以“龙胜梯田”概念打造的叠水荷池、小桥叠水，加上风情泳池等等，构筑动静相宜、声色光影水体景观。

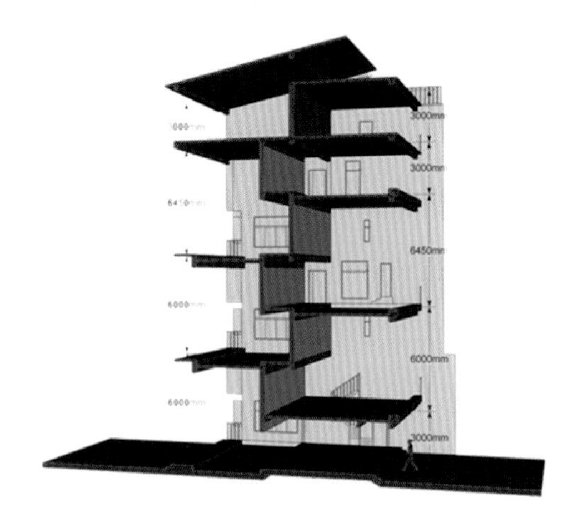

错层式花园剖面 Section of Splitlevel Garden

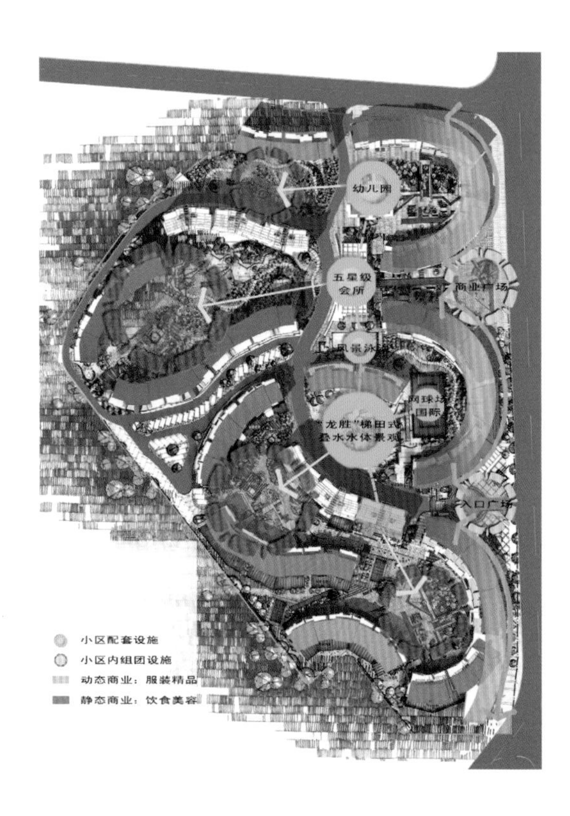

Living Space System

To avoid the retained original 138 trees, it builds arc-shaped residential buildings that have difficulty both in design and construction. Arc corner various from 15 degrees, 25 degrees, 30 degrees, 40 degrees to 45 degrees to protect each tree as much as possible. In ordinary circumstances, distance between multi-layer residence is about 20~30 m, while that between small high-rise is about 30~40 m, which ensures more than 80 percent of households to obtain 60~120 landscape sight distance. Free-form curve design with low wind drag coefficient, two households on each floor share one elevator, completely hollow ventilation design of hanging gardens, all these enjoy excellent ventilation condition. 6m high hanging gardens for each household narrow the distance between you and nature each moment. Split-level interior space creates warm home life and enhances the privacy. Splendid sunshine shines the kitchen and bathroom, improving the humid environment and reducing mosquito breeding. The trumpet of the building layout naturally formed the flared welcome gesture in the entrance plaza and commercial plaza, and effectively blocked the noise emission from the east side while extending the commercial interface.

Supporting Service System

Supporting facilities planning strengthens the good interaction between facilities and facilities and is conducive to realize the nice transition from public space to private space. Longsheng terrace courtyard in the center links with swimming pool and tennis court, forming a seamless exchange space. The club is an extension of the living room. The leisure club, swimming pool and several aerial leisure spaces provide indoors, half door and outdoor shared communication space suited to the needs of different groups. Diversity: 15,000 m^2 of commercial facilities, star nursery, golf greens, stylish swimming pool, international standard tennis courts, table tennis, chess and other supporting facilities.

Road Traffic System

Pedestrian traffic design: first to meet convenience needs of necessary activities: commuting, shopping, work and travel; then selective activities—walking, playing, strolling; and then social activities: conversation, communication, meetings, etc. and the last to meet the need of landscape and atmosphere. Driveway design slows the vehicle speed through the use of curved shape, links two exits in the south and north and serves each group efficiently. Garage configuration: the size of the project is just over 68,000 m^2 with a total construction area of 116,000 m^2. In order to protect the original trees, it has to set 7 garages. It is commendable that all of them are semi-underground with natural ventilation and lighting.

Landscape Environment System

Woodland boasts a tree-lined walkway over 1500m in the 68,000 m^2 area. Walking in the Finely crafted long path, you can catch the different views by taking each step. 40-year old tall native trees accompany with you and cheer you up like a piece of beautiful music. Facade landscape—asymmetric design, varied framework like jumping notes, free-curve construction interface design—obtains new image every day with the changes of season, light and shadow. Multi-dimensional greening: well-designed vertical and horizontal greening with the green hanging garden of each household, 40-year old trees and new sprout build a variety of multi-dimensional green landscape. No water no habitants: Lotus pond, bridge, swimming pool, etc. constructed after the concept of "Longsheng Terrace" create both steady and dynamic and colorful waterscapes.

A RETREAT AWAY FROM THE WORLD

Taohuayuan Golf Resort, Changde, Hunan

清净 自然 尊贵 优雅—— 湖南常德桃花源高尔夫会所

项目地点： 中国湖南省常德市
室内设计： 刘波设计顾问（香港）有限公司
设 计 师： 刘波

Location: Changde, Hunan, China
Interior Design: Paul Liu Design Consultants (HK) Co., Ltd.
Designer: Paul Liu

本项目位于湖南省常德市柳叶湖畔，作为天利桃花源国际高尔夫俱乐部的重要组成部分，愿景是打造一个自成一体的休闲世界，让来此休闲度假的宾客能够在享受高尔夫运动快乐的同时，享受到体贴尊贵，卸下多时疲惫，在这个如世外桃源般清净、又有着皇家行宫气派的美好天地里放松身心。

整个会所内设施齐全，配备有酒吧、自助餐厅、休息室、娱乐室、水吧以及VIP会议室、自助银行和各式景观露台等，可以满足不同的需求。各个部分之间联系密切，让来此的客人有种宾至如归的感觉。

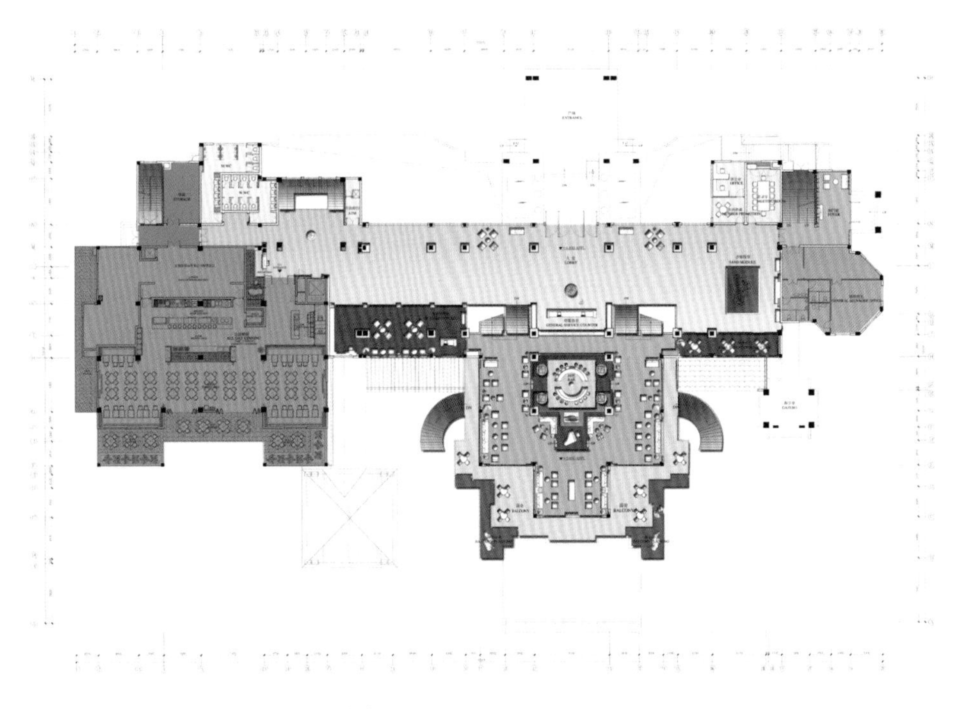

大堂平面图 Lobby Floor Plan

室内设计上，设计师创造了拥有优越条件的大堂建筑空间，体现大气磅礴之感；室内设计均采用自然气息浓郁的材质：种类各异的云石、自然色泽的木材、藤编家私、真丝窗帘、丝绒及麻质布艺。

室内装饰手法浪漫质朴，线条磊落；地面用自由随意的云石进行混搭碎拼，墙面素雅利落，天花沿用其原本建筑形状用传统手法进行美化，塑造一场豪气又质朴的感官享受。

整个室内氛围亦与周围尘世相隔、清灵、优雅的气息相互映衬，使得会所的活动空间内外能够享受阳光沐浴、微风拂面、美景拥抱以及获得清新怡人的空气，充分体现自然、和谐之美的设计理念。

Taohuayuan Golf Resort, located by the Liuye Lake of Changde, Hunan, is an earthly paradise isolated from the crowded outside world which lets you enjoy the pleasure of golfing yet at the same time relax and enjoy the royal treatment in this tranquil paradise.

The club is fully equipped with bar, cafeteria, lounge, recreation room, water bar, VIP meeting room, Self-Service Banking and a variety of landscape terraces for different requirements. All parts connect with each other closely to make the guests feel at home.

In terms of interior design, the lobby is spacious hence designed with natural materials including a variety of marbles, timbers, furniture made of vines, silk, velvet and linen curtains.

The decoration style is romantic yet modest. The floor is covered with shattered and mixed marbles, while the wall and the ceiling are decorated to show pure beauty and grandness.

While the project is surrounded by natural aroma outside, it is also full of harmony inside. You can get bathed with sunshine and breeze, surrounded by fresh air and pretty scenes, natural and harmonious beauty.

DUPLEX BUILDING WITH WALLS ON THREE SIDES AND AN OPENING VIEW THE OTHER

Two Family House in Kaltern

三面围墙　一面曙光的复式建筑——卡尔顿双户住宅

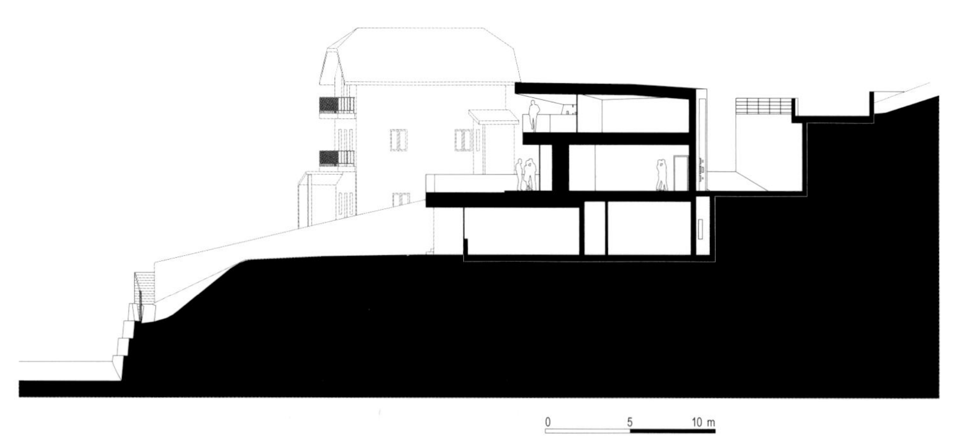

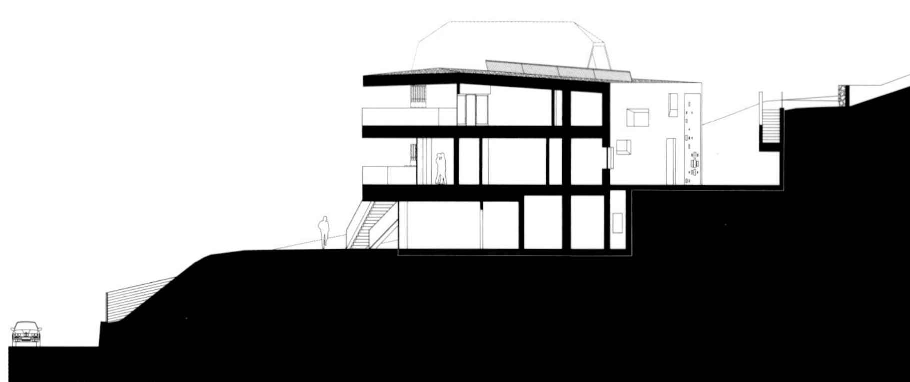

横切面图 Cross Section

项目地点： 意大利波尔扎诺卡尔顿
客　　户： Giacomuzzi
建筑设计： 意大利Monovolume建筑师事务所
建筑面积： 600 m²
摄　　影： Ren é Riller

Location: Kaltern, Bolzano, Italy
Client: Giacomuzzi
Architectural Design: Monovolume Architecture + Design
Building Area: 600 m²
Photography: Ren é Riller

项目概况

Giacomuzzi双户住宅是一个复式建筑，高度紧凑的体量不仅解决了能源问题，还顺应地形，充分利用了周边景观。它位于Caldaro sulla strada del vino村庄的郊外，四时向阳，田野和葡萄园从这里一直延伸到Caldaro河畔，一片开阔景致。

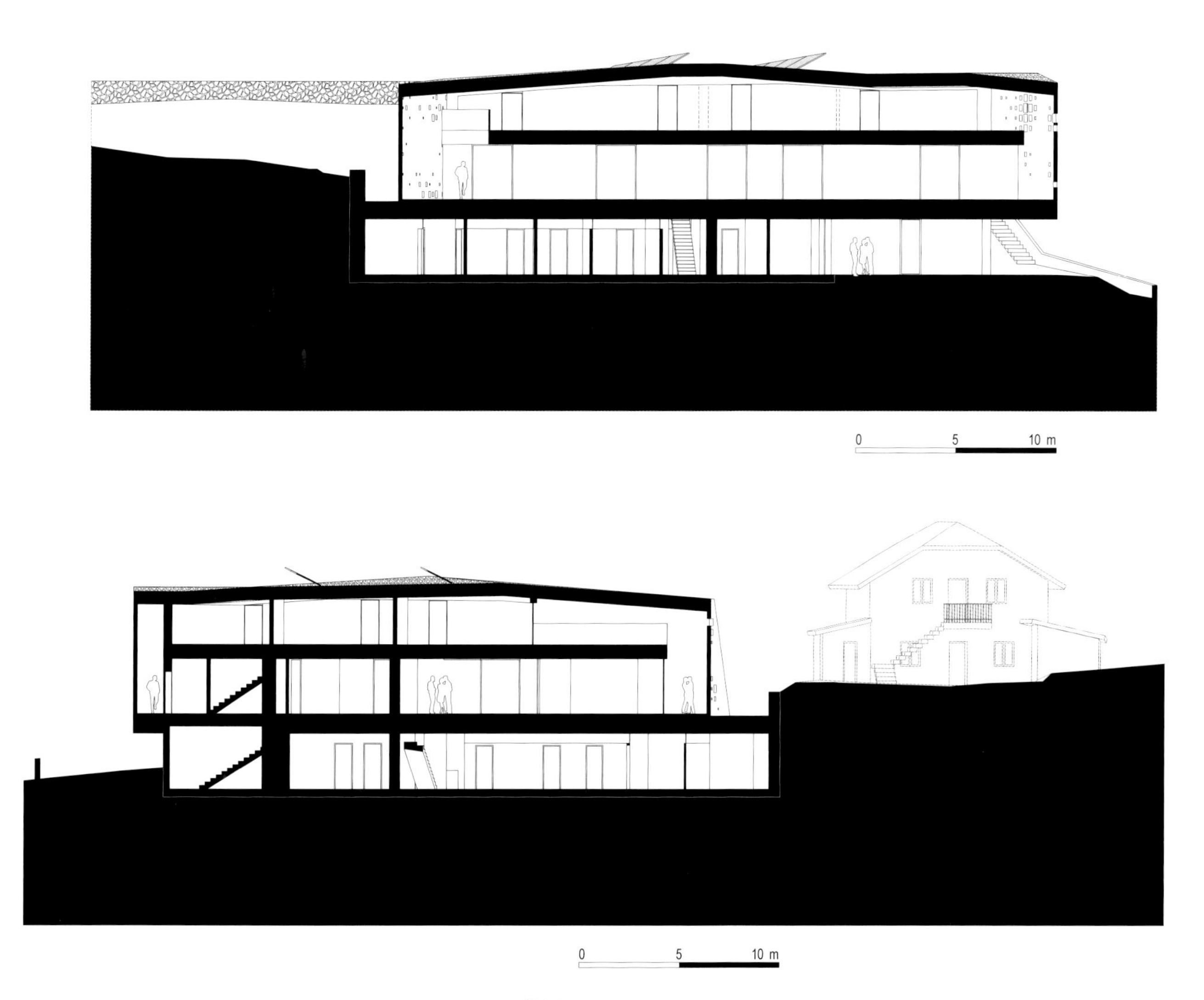

纵剖面图 Longitudinal Section

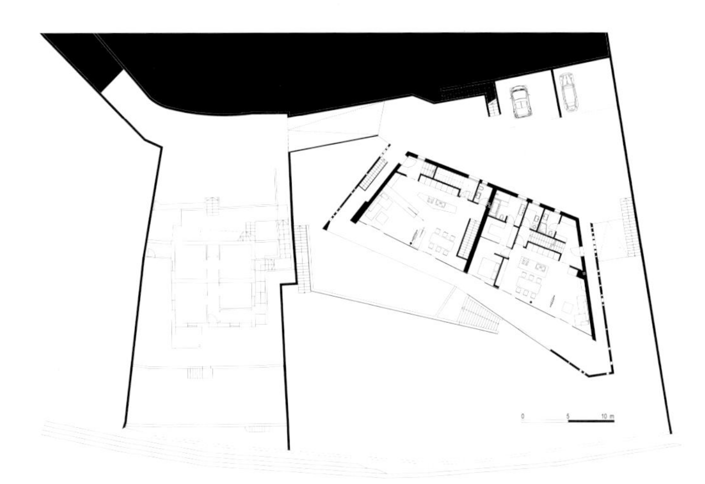

底层平面图 Ground Floor Plan

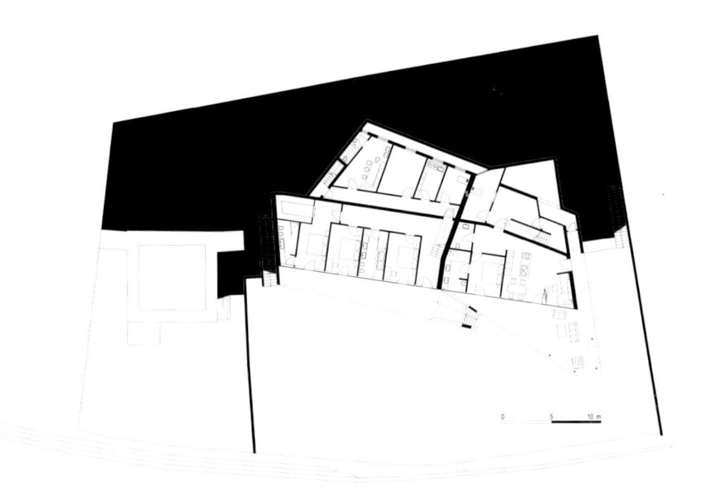

地下层平面图 Basement Plan

规划布局与建筑设计

Giacomuzzi双户住宅三面围墙，唯独南面设有大窗户和阳台，就是为了不失去这一美景。客厅和卧室朝向南方，其它房间均朝西北。这类复式住宅对传统的坡顶房屋进行了现代意义上的诠释，如利用太阳能。整座房子为了顺应天然斜坡，较低的立面走势稍有变化，并不单一。

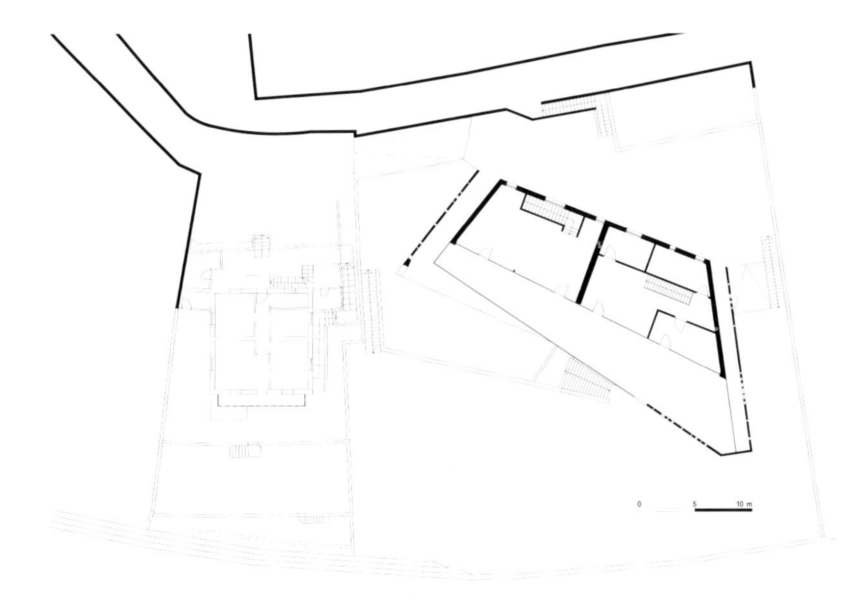

一层平面图 First Floor Plan

Profile

Giacomuzzi house is a duplex building. It is conceived as a highly compact body to fulfill energy issues and it fits into the sloping terrain to exploit the landscape. The house is oriented towards the path of the sun and overlooks a stunning sight. It is located in the outskirt of the village Caldaro sula strada del vino, from where you could enjoy the view as far as to Lake Caldaro with only filed and vineyards around the house.

Planning Layout & Architectural Design

To show up this wonderful view, a continuous wall closes three sides of Giacomuzzi house, whereas the south facade features wide windows and balconies. Living room and bedrooms face to south, the other rooms to northwest. This type of duplex house is the modern interpretation of the traditional pitched roof houses, to exploit the solar energy. Following the path of the natural slope, the position of the first and second floor is shifted backward from the lower floor.

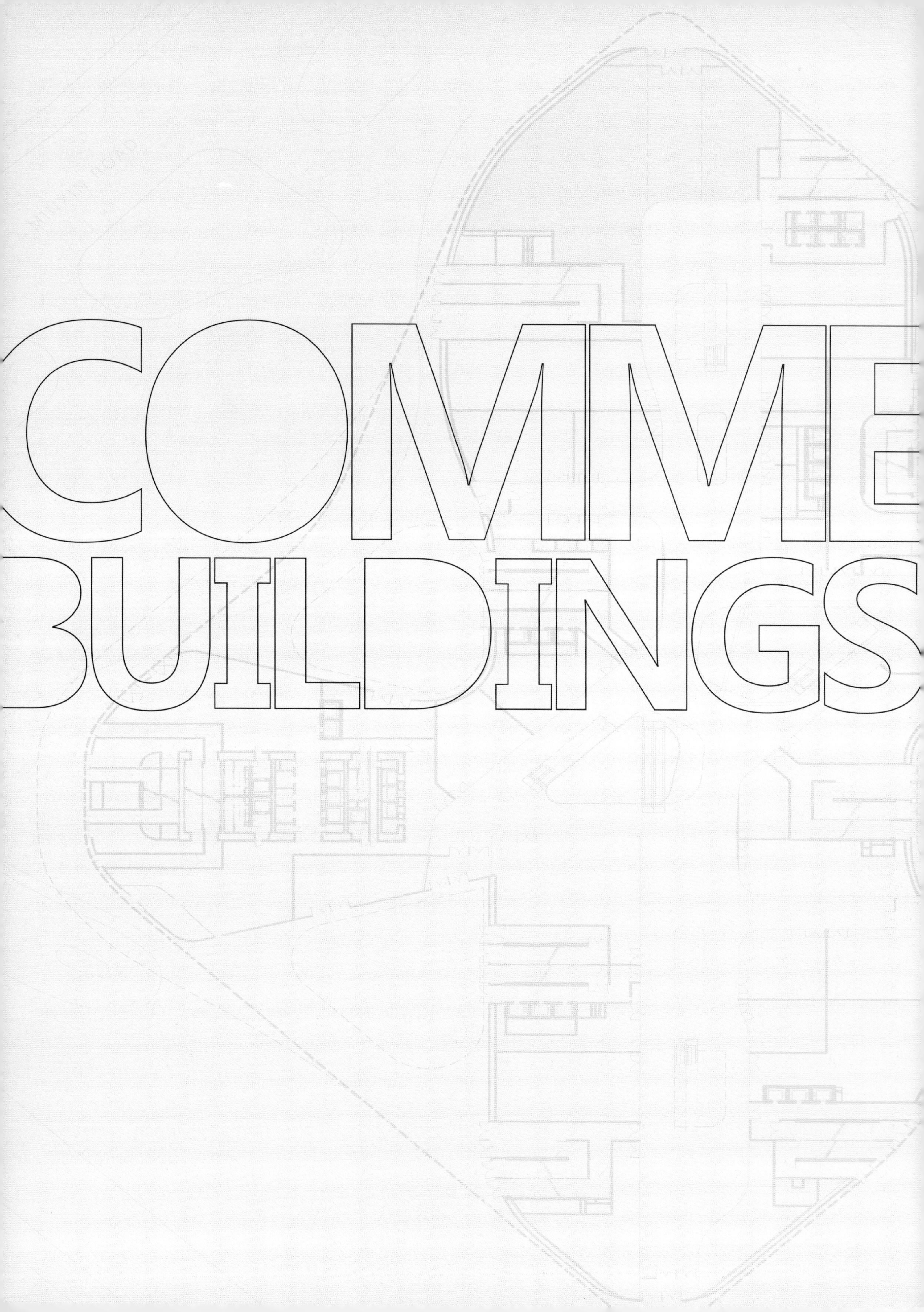
COMME
BUILDINGS

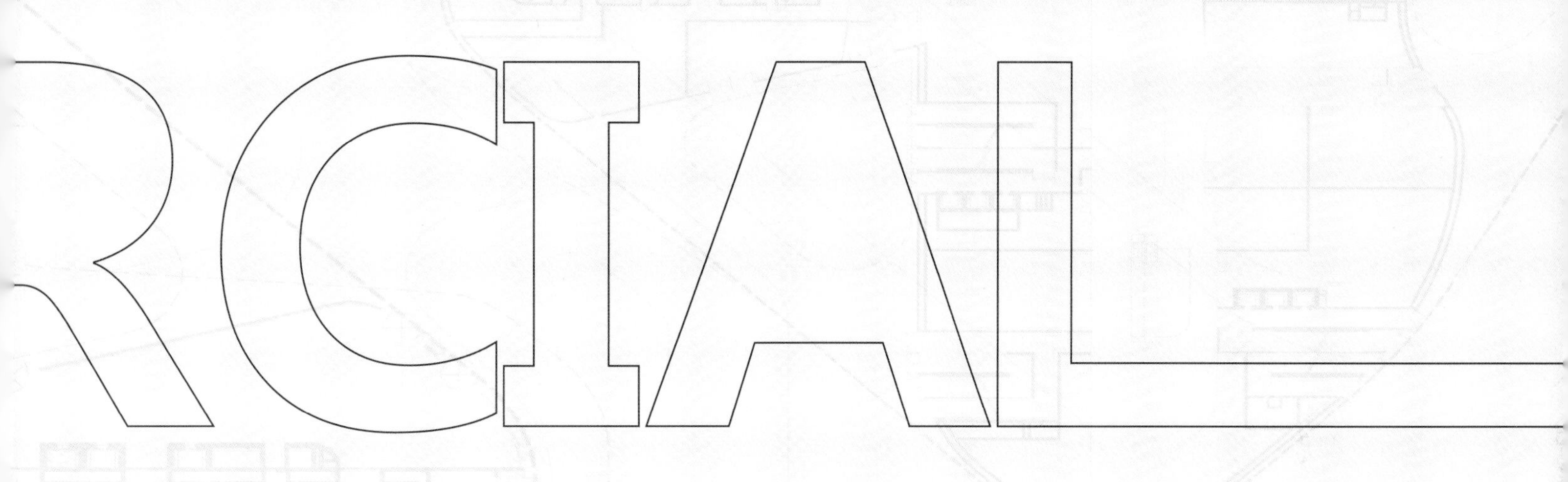

商业地产

林荫道广场：
神秘而典雅的弓状穆斯林特色建筑

新燕赵国际商务中心3S综合体：
将商业空间城市空间化

那不勒斯购物中心：
清晰可见的内部展现同质形象

MYSTIC AND ELEGANT MODERN ISLAMIC ARCHITECTURE| Boulevard Plaza

神秘而典雅的弓状穆斯林特色建筑

—— 林荫道广场

项目地点：阿拉伯联合酋长国迪拜
客　　户：艾马尔地产公司
建筑设计：凯达环球有限公司
占地面积：17 200 m^2
建筑面积：60 927 m^2

Location: Dubai, UAE
Client: Emaar Properties PJSC
Architectural Design: Aedas Limited
Site Area: 17,200 m^2
Floor Area: 60,927 m^2

项目概况

Boulevard Plaza的两座塔楼位于通往迪拜塔的街道上，塔高为173m，因为如此靠近街对面的世界最高塔——迪拜塔而格外引人瞩目。

规划布局

两座塔楼均面向主入口，向来访人员展示出欢迎的姿态。作为该区开发的延续，Boulevard Plaza塔楼微微扭转，向街对面的迪拜高塔致以崇高的敬意。两座塔楼分别高42层和34层，用作甲级写字楼。

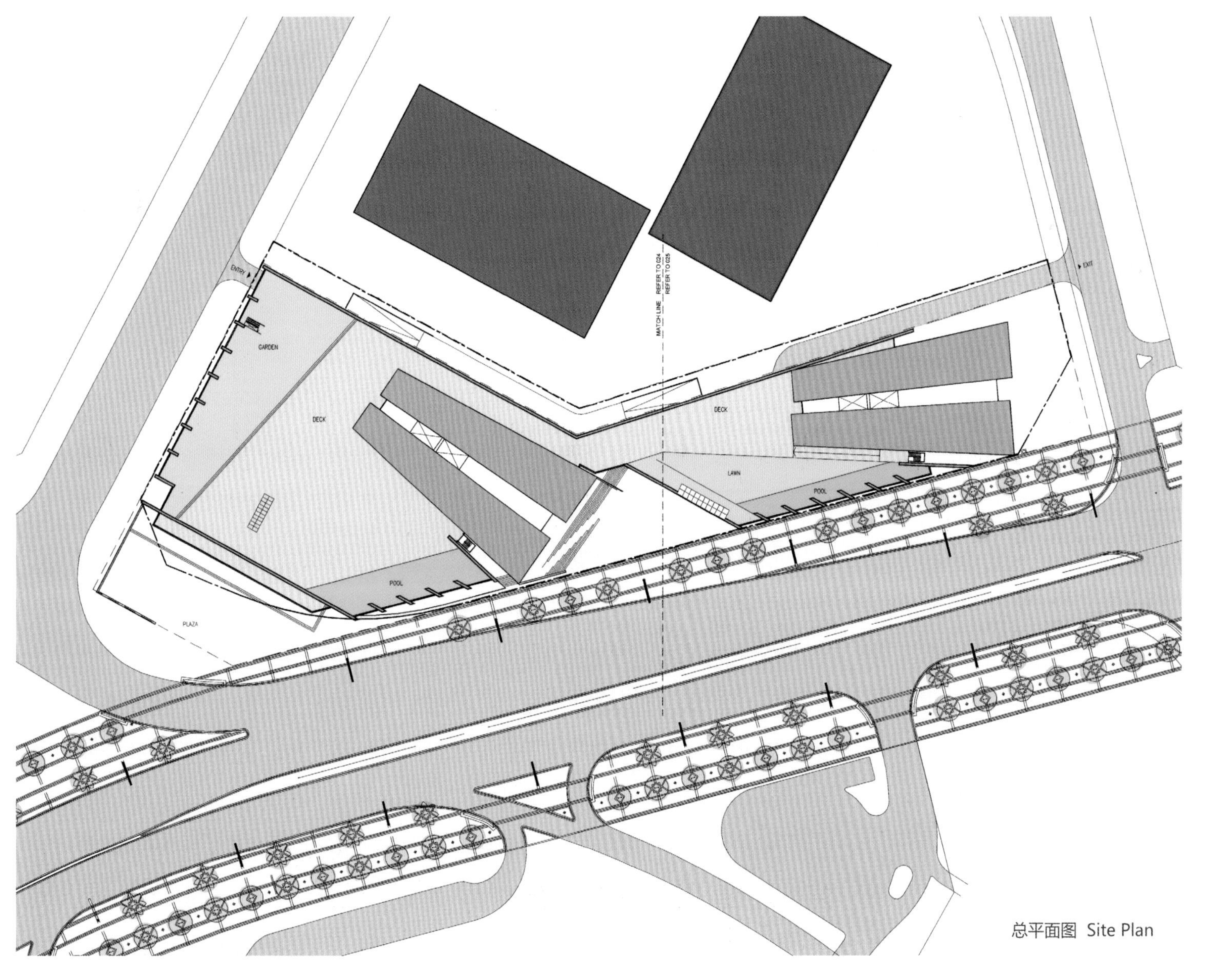

总平面图 Site Plan

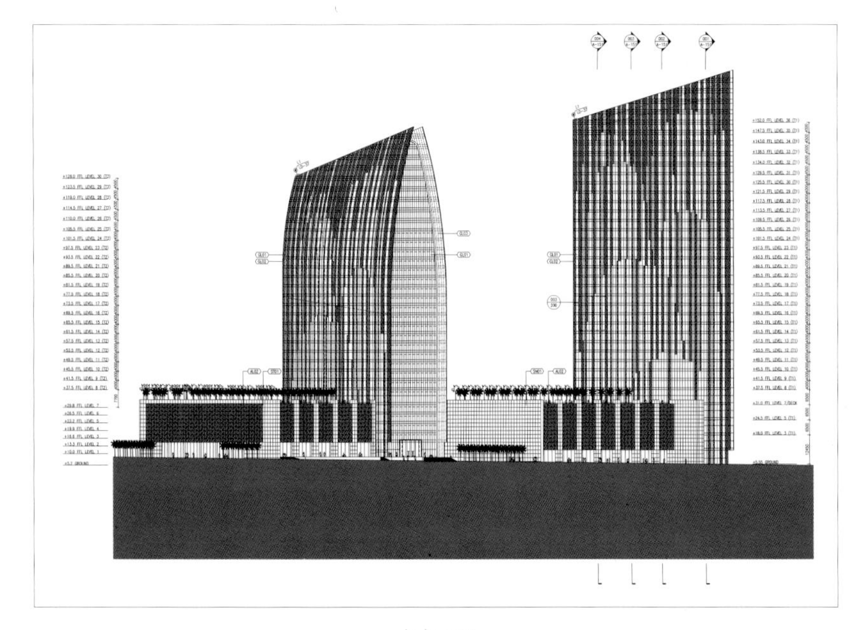

南立面图
South Elevation

建筑设计

设计旨在使Boulevard Plaza与迪拜塔完美融合，表达对该世界最高塔的崇敬之情。塔楼的形式及结构源于对周围环境的考虑及其作为一座现代化穆斯林建筑的象征意义。它将恰如其分地融入这座世界上最现代化的穆斯林城市——迪拜。

塔楼表皮极富传统穆斯林建筑特色，使人联想到面纱。随着楼层上升，两侧向中间弯曲，形成弓状，既令人产生敬畏感，又显得高贵典雅。尽管拥有曲线变化的形体，建筑的各个部分却通过模块化的组合，形成了简洁的结构和理性的空间，同时有效地控制了成本。

具有现代穆斯林特色的立面图案使之与周围环境融为一体。同时特殊的材料使之成为性能良好的遮阳板，使建筑在迪拜强烈的日照下，有效地降低热负

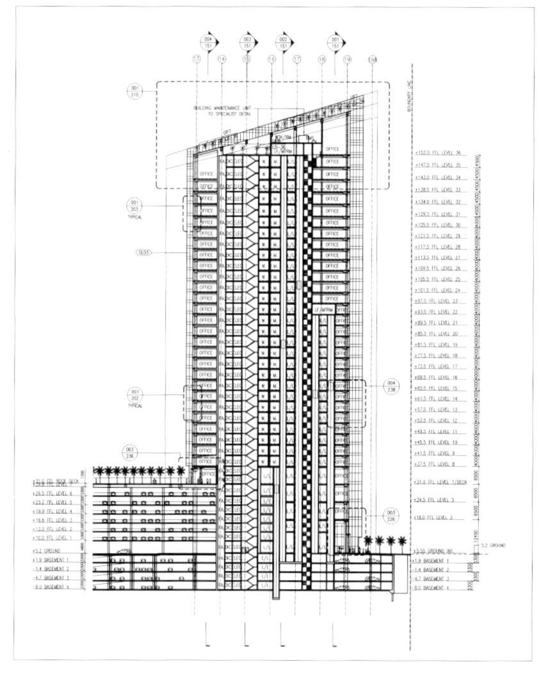

纵剖面图
Longitudinal Section

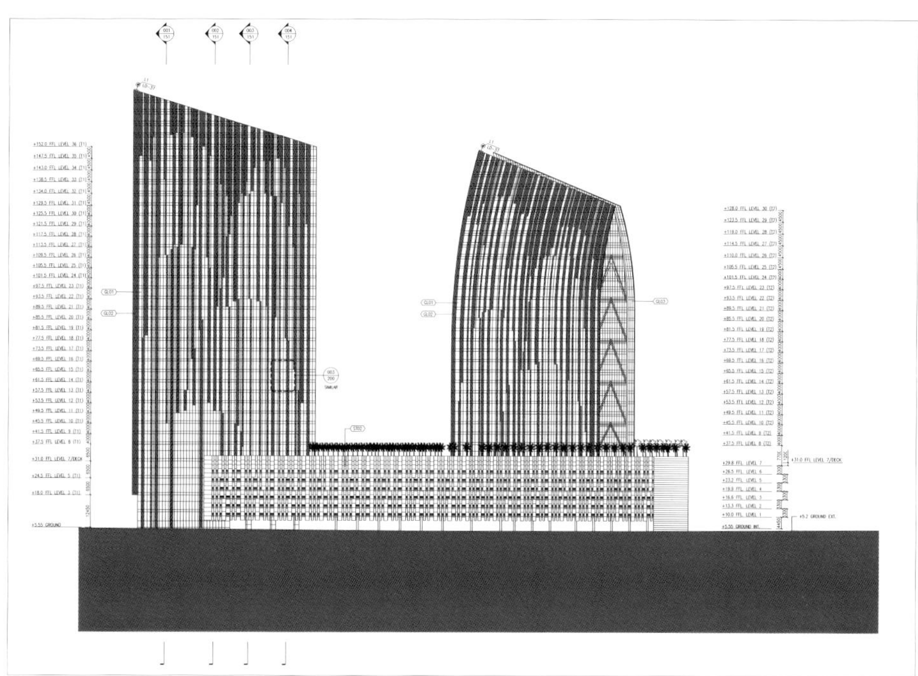

北立面图
North Elevation

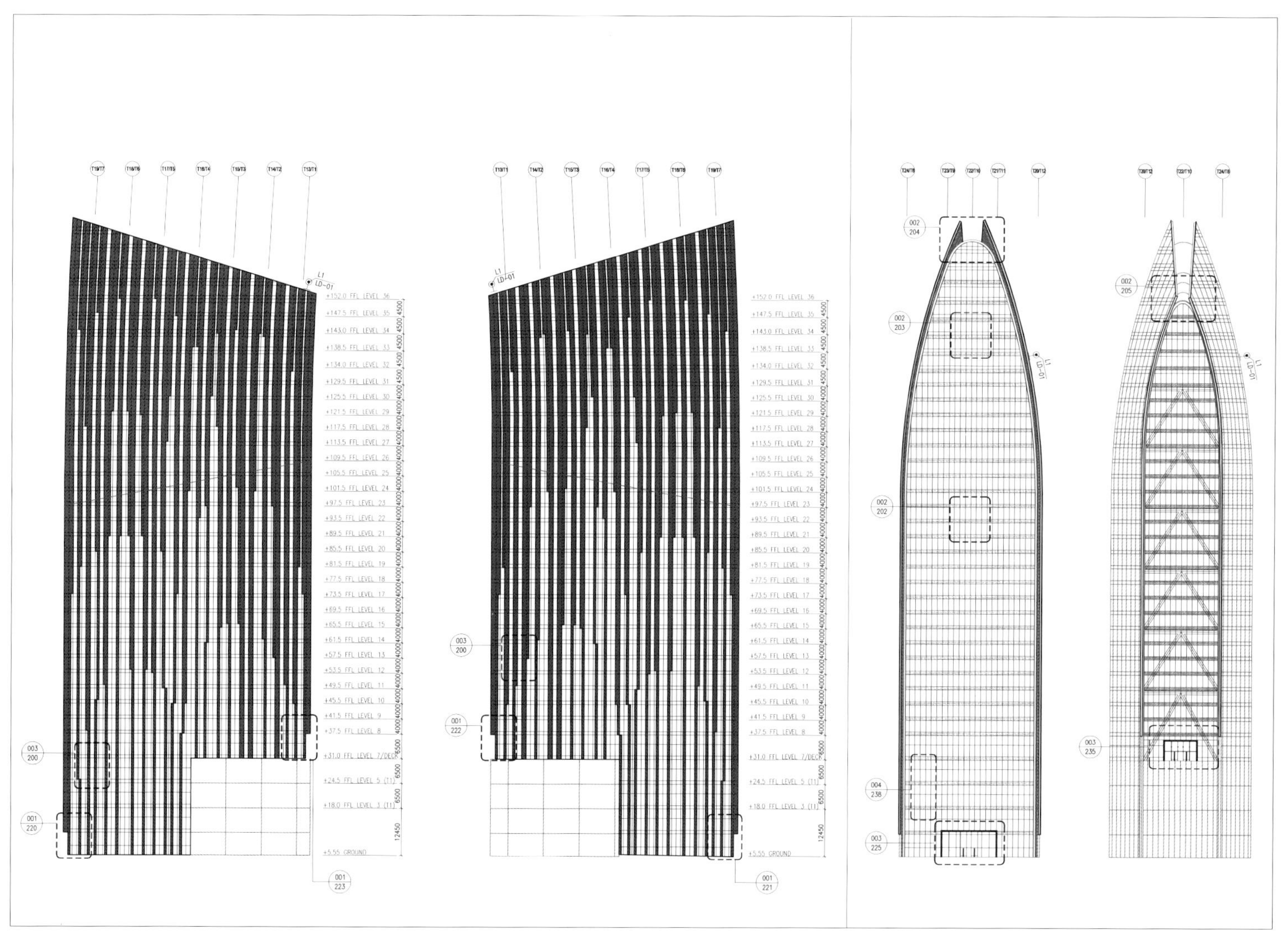

一号楼北立面图
North Elevation Tower One

一号楼南立面图
South Elevation Tower One

一号楼东立面图
East Elevation Tower One

一号楼西立面图
West Elevation Tower One

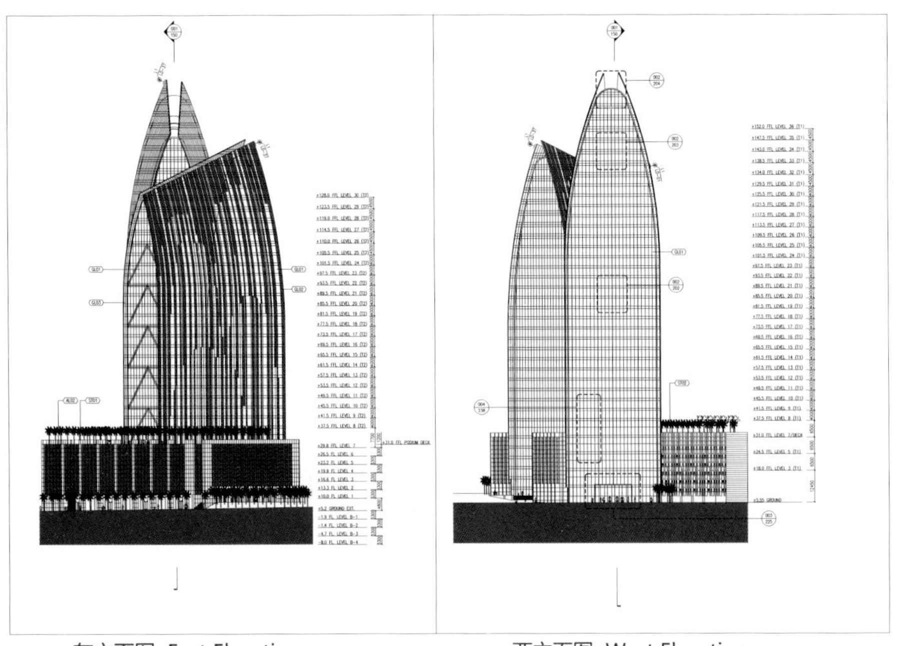
东立面图 East Elevation　　西立面图 West Elevation

荷，从而减少了机械降温的成本。突出的立面向外延伸5m长，为采用透明玻璃的东西侧立面遮挡阳光；而南北立面则采用绘有图案的玻璃。

裙楼是一座露天结构，采用自然通风和风扇通风。在开放的立面上，巨大的柱廊间镶嵌着绘满图案的金属屏，尽显穆斯林特色。

设计考虑到结构的优化，将剪力墙设置在靠近外走廊的一侧。这样一来就拓宽了基础结构，减小了核心与立面之间的距离。同时这种设计也有效地缩短了建筑进深，节省了建筑材料，从而降低了能耗。扩大的核心结构将压力转向地基部分，节省了混凝土的用量。另一个十分重要的降温措施在于屋顶景观区的水景设计。该设计有效地达到了自然降温的效果。而屋顶的软景观同样起到了降温的作用。

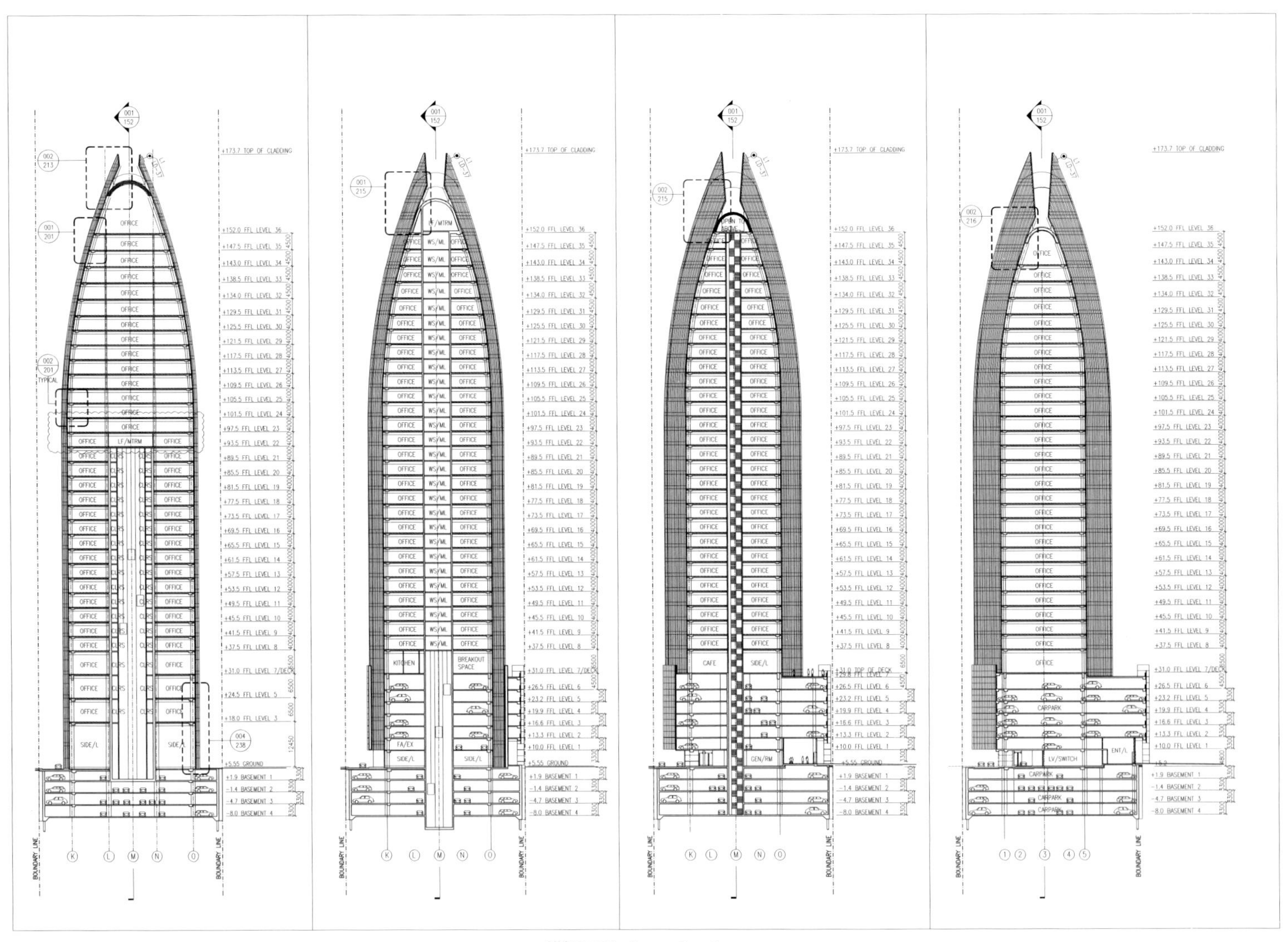
横切面图 Cross Section

Profile

The Boulevard Plaza Towers, which are high to 173m, stand at the gateway into the Burj Dubai development. The importance of the site is accentuated even more by being located immediately across the street from the tallest tower in the world – Burj Dubai Tower.

Planning

Both towers point toward the main entry to greet the visitors. As one continues into the site, the towers rotate their orientation as a gesture of respect to the lofty neighbour across the street. The two towers of 42 and 34 floors contain grade A+ office space, looking out to take advantage of the views toward and around the Burj Dubai.

Architectural Design

The design strives to fit appropriately into this development as a respectful icon for the community. The relationship of the forms and their articulation derive from both a contextual response and the building's

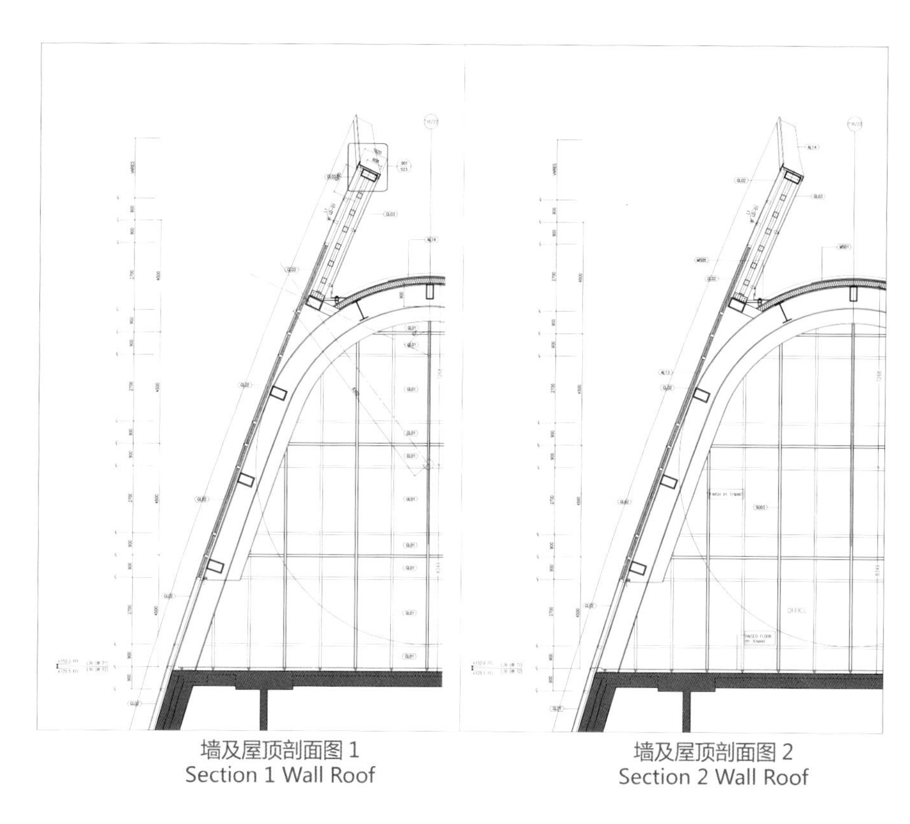

墙及屋顶剖面图 1
Section 1 Wall Roof

墙及屋顶剖面图 2
Section 2 Wall Roof

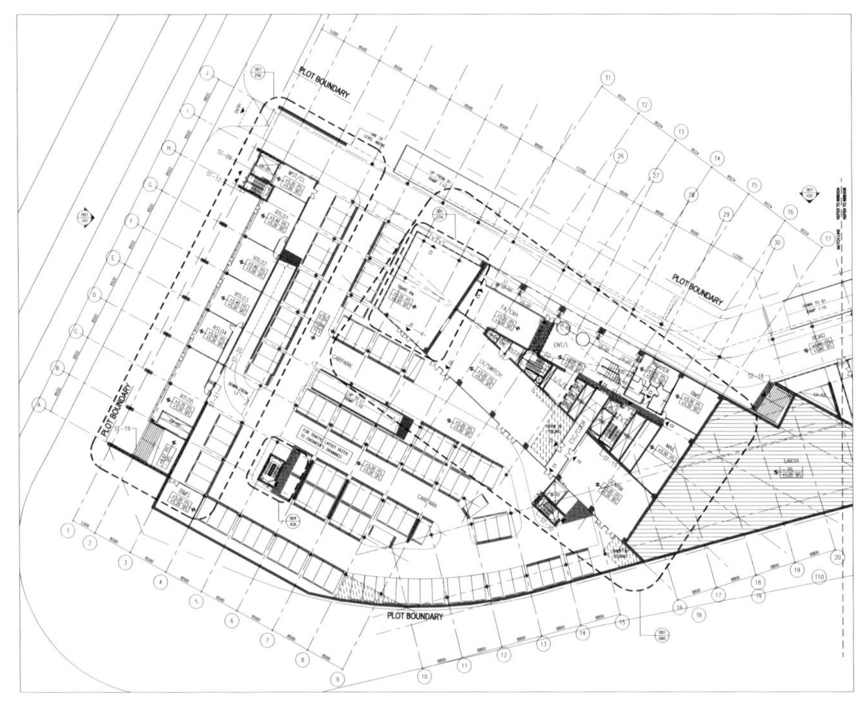

底层平面图 Ground Floor Plan

symbol representing a modern Islamic architecture set appropriately within the most modern Islamic city in the world - Dubai.

The towers are clothed with an articulated skin recalling the veils and layers of traditional Islamic architecture. As the figures rise, they bend inwards, forming two deep, shadowed arches up to the sky and beyond - toward the top of the Burj Dubai. Despite the changing curvaceous form of the building section, the units are modularized to standard layouts for construction simplicity, rational space and cost efficiency.

While the modern Islamic feature facade patterns offer symbolism to address its context it also acts as a sun screen which significantly reduces heat loads applied by the intense Dubai sunlight thus reducing energy consumption produced by mechanical loads. Over-sailing facades are cantilevering up to 5meters length offering shade to the East and West elevations which contain a more transparent glass than the North and South patterned facades.

The Podium is an open-air structure with natural ventilations and fans for air flow circulation. The open facade is clothed with patterned metallic screens in between a monolithic colonnade that recalls the Islamic motif.

The design considers the structural efficiency by pushing shear walls to the outside of the corridor, therefore widening the structural base and reducing the span between the core and the facade. This effectively reduces the structural depth and construction materials required for the structural members, thereby reducing embedded energy consumed. The larger core allows pressure to be transferred to the foundation over a larger footprint thus reducing quantities of concrete. Another important heat reduction measure is the water features along roof landscaped areas, which allows natural cooling for occupants. The soft landscape on roof surfaces also minimizes solar gain.

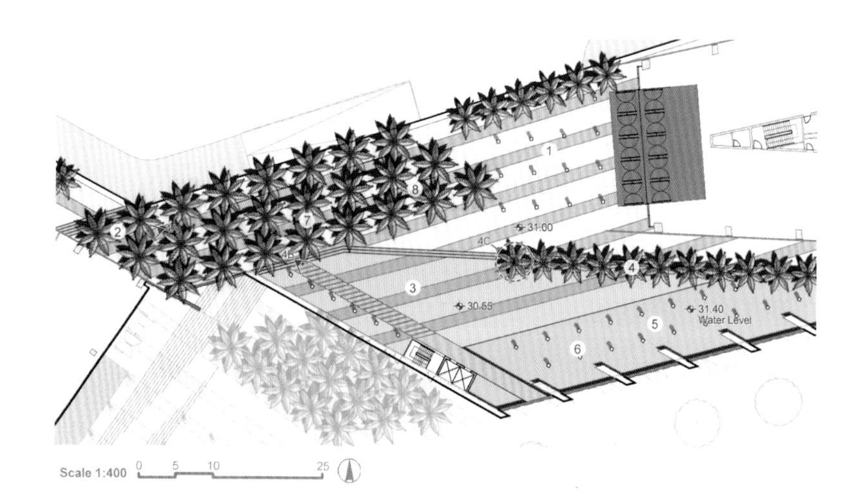

4A: View of Outdoor Cafe Seating Area

4B: View of Open Lawn Area

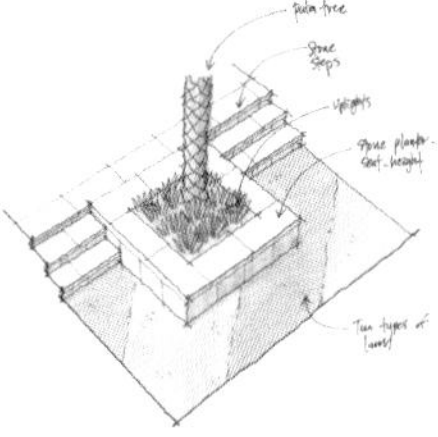
4C: Detail of Planter & Steps

TURNING COMMERCIAL SPACE TO PUBLIC URBAN SPACE

| 3S Complex of New Yanzhao International Business Center

将商业空间城市空间化 —— 新燕赵国际商务中心3S综合体

项目地点：中国河北省保定市
建筑设计：广州市金冕建筑设计有限公司
总用地面积：11343 m^2
总建筑面积：113205 m^2
计容建筑面积：90394 m^2
容积率：7.97

Location: Baoding, Hebei, China
Architectural Design: King Made Group
Total Land Area: 11,343 m^2
Total Floor Area: 113,205 m^2
Floor Area (take into account in plot ratio) : 90,394 m^2
Plot Ratio: 7.97

项目概况

新燕赵国际商务中心项目位于河北省保定市，具备得天独厚的区位优势。项目定位为“3S”——SKY写字楼+SOHO公寓+SHOPPONGMAL商业的现代城市综合体。

规划布局

项目针对周边建筑环境，创造高层高、大空间、高高度的SKY写字楼，既弥补了市场空间，同时又提高了整个项目的档次。项目设计是以小型SOHO公寓满足大量小型企业和商务人士市场需求，不仅利于资金快速回笼，也避开传统公寓市场竞争。这样实现项目地块的价值最大化，以SHOPPINGMALL的形式获取更多的商业面积，有利于与周边的传统百货形成差异化竞争。另外，商业人流主要设置在城市主干道的沿街界面，而在路口处设置主商场，可利于举行大型商业活动。

建筑设计

建筑形态设计以“将商业空间城市空间化”为目标，创造巨大的空中花园，与城市公共空间相映成趣，提升项目开放度，空中花园上空是某跨国企业总部办公室。商业裙楼顶以及空中花园种植绿化，打造三维绿化体系；建筑以钢构、玻璃、石材为主材料，以方、圆等基本形体，加以斜线创造动感立体的建筑形体，营造都市中心氛围。

写字楼设计突出“SKY”理念——配置独立、专用观光电梯筒；针对企业总部办公室需求设置在80~100 m高空位置；当中也设置了4m层高，局部有6 m或5 m层高；配置了共享的休闲绿化通天中庭。

公寓突出“SOHO”功能需求——每两层为一个单元，共享6.6 m高的休闲、会议、聚会空间，并靠近共享空间设计面积较大的中高档“SOHO”公寓；设置开放式厨房，可商可住；面积以60 m^2左右为一户，辅助少量80 m^2左右的户型，可随意合并和打通2套或3套；为酒店式管理预留空间和配套设备；层高3.3 m，避免了梁板过于低矮，局部4 m层高。

商业以“SHOPPINGMALL”形式内容为指导——强调以休闲、娱乐、餐饮、购物为一体，具有统一的管理理念和实体的购物中心；内部空间设计强调“街道”的理念，以“L”型中庭创造室外感觉的步行街空间，再辅以小型内街，提高了商业铺面价值；设计以经典的“十”字型主要交通通道加“回”字型内接组织交通，实现了“无商业铺面死角”设计，充分发挥负一层商铺的高商业价值的特性，尽可能创造更多的负一层商业面积，如可达到了近5 000 m^2的面积。

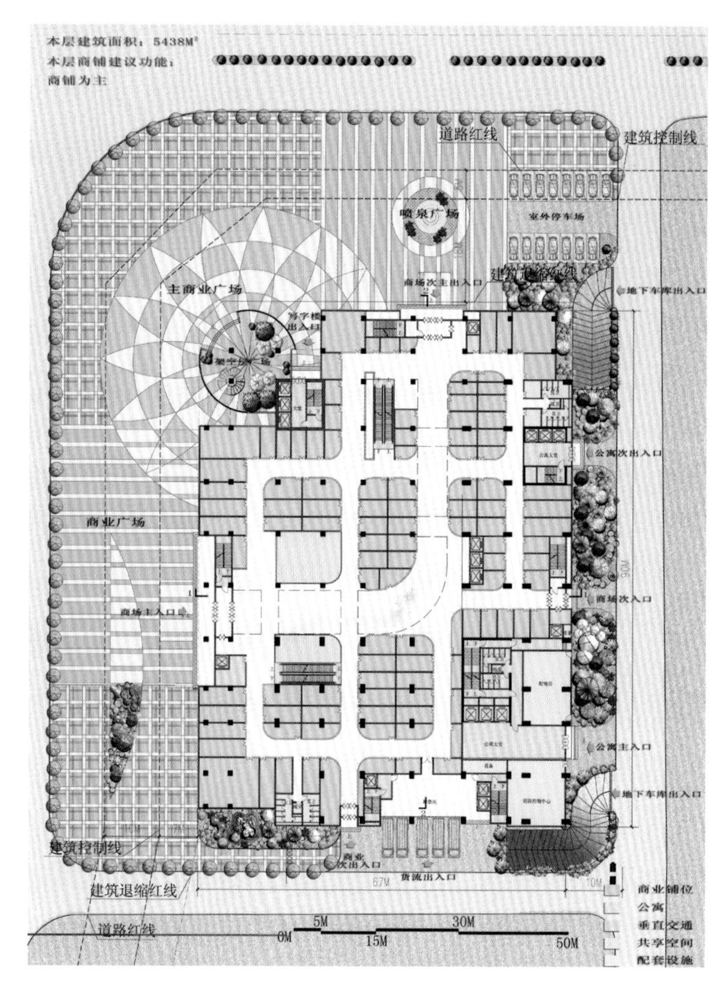

首层平面图 First Floor Plan

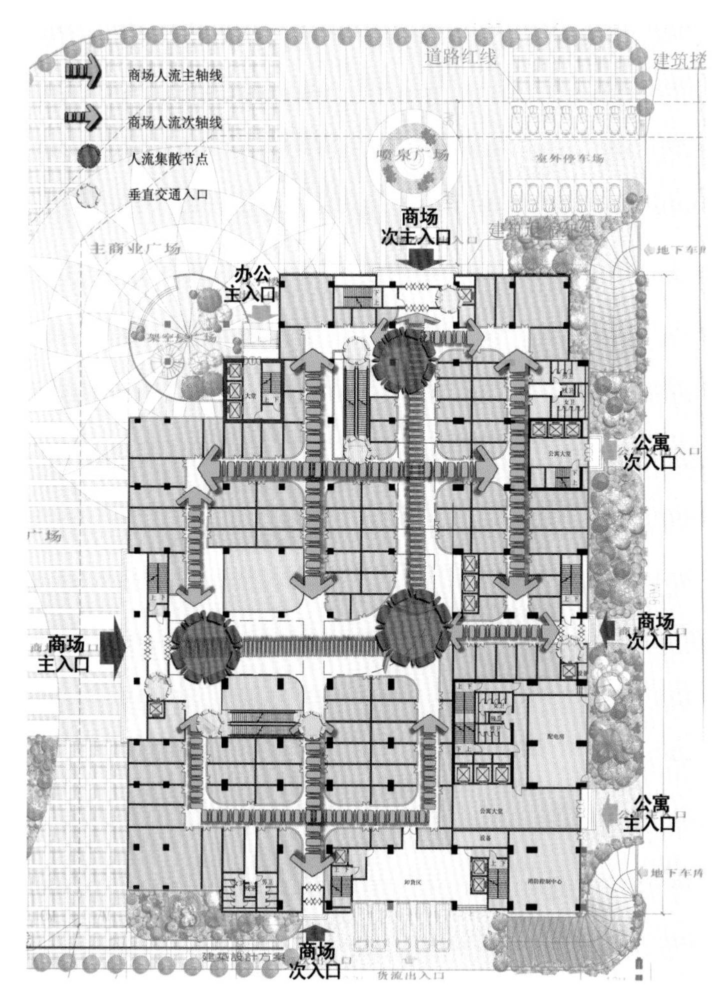

商场人流分析图 Commercial Streamline Analysis

Profile

Standing in Baoding, Hebei, the project features an advantaged location. It is envisioned to be a 3S urban complex – Sky Office, SOHO Apartment and Shopping Mall.

Planning

Considering the surrounding architectural environment, the Sky office building is designed to be tall with great floor height and large space. It not only makes up the office market but also promotes the grade of the whole project. Small SOHO apartments will meet the requirements for small companies and business men. It benefits will provide rapid withdrawal of funds and avoid the competition with traditional apartments. In addition, Shopping Mall is designed to get more commercial value and shape difference from the surrounding traditional department stores. What's more, the commercial flow is arranged along the street. With the shopping mall at the intersection, it is easy to hold large-scale commercial activities.

Architectural Design

With the objective of "turning commercial space to be part of the urban space", it has created a huge sky garden which echoes the public urban space and enhances the openness of the project. Above the sky garden, it is the headquarters office of a multinational corporation. The top of the commercial podium and the sky gardens of the tower buildings are set with green plants to establish a three-dimensional green system. Main building materials include

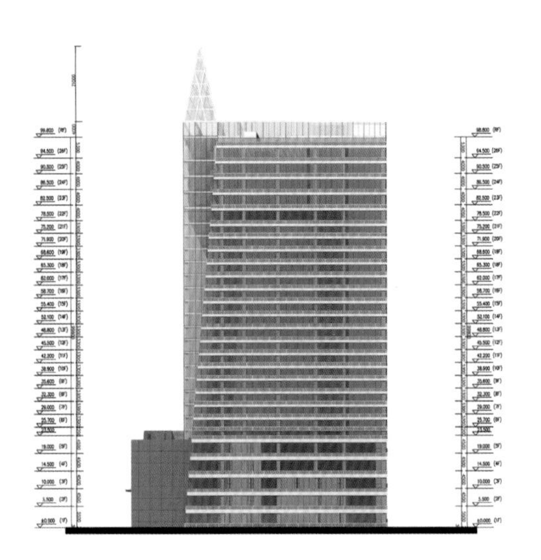

南立面图 South Elevation

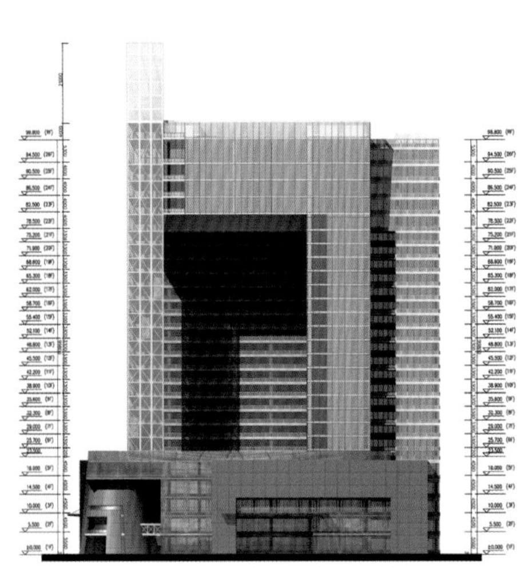

西立面图 West Elevation

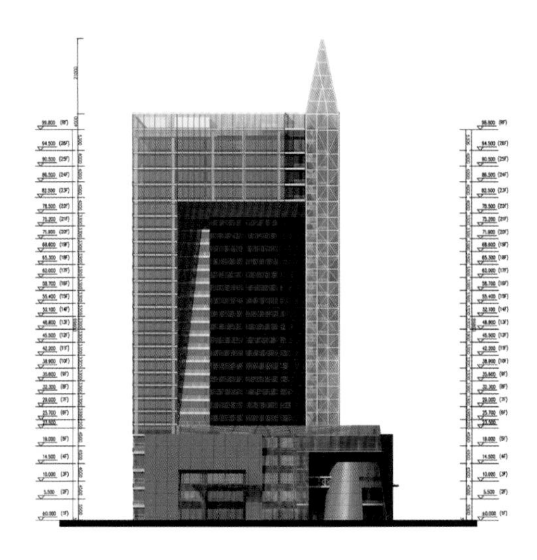

北立面图 North Elevation

steel, glass and stone. Square and round are the basic shapes which form dynamic and dimensional building forms together with oblique lines, creating strong downtown ambiance.v

The office building is designed to highlight the idea of "SKY" with independent sightseeing elevator. Headquarter offices are set 80~100m high with a floor height of 4m, 6m or 5m. There is also the open atrium for public relaxation.

The apartment building is designed to highlight "SOHO" function. Every two floors share a 6.6m high space for leisure, meeting and party. And near to the public space, high-end SOHO apartments are set around with open-style kitchen for both business and living. Most of the units have a floor area of 60m^2, and some have 80m^2. They can be combined flexibly by two or three. Rooms (3.3m or 4m high) and equipment are left for future hotel-type management.

The commercial part is presented in the form of Shopping Mall to emphasize the integration of leisure, entertainment, catering and shopping. It is a shopping mall that combines managing idea with the building itself perfectly. The interior spaces are arranged along a "street". L-shaped atrium gives people the feeling of an outdoor pedestrian street. Additionally, there are also many small interior streets to increase the commercial values. Classical “十” shape and “回” shape are used to organize the traffic system and avoid dead corners. Thus the commercial value of the basement one floor is maximized with a floor area of about 5,000m^2.

HOMOGENOUS IMAGE INSIDE THE CITY | Shopping Center, Naples

清晰可见的内部展现同质形象—— 那不勒斯购物中心

项目地点：意大利那不勒斯
建筑设计：法国Silvio d'Ascia建筑师事务所
合作设计：TECNOSISTEM SpA (P.M. et D.L. Ing. Marco Damonte)
摄　　影：Barbara Jodice

Location: Naples, Italy
Architectural Design: Silvio d'Ascia – Architecte
Partner: TECNOSISTEM SpA (P.M. et D.L. Ing. Marco Damonte)
Photography: Barbara Jodice

项目概况

该购物中心是那不勒斯邮局改造项目的一部分，靠近Arenaccia住宅区。它位于那不勒斯中央火车站的北部，车站与机场之间。它的出现满足了城市住宅区对休闲设施、商店和城市再度认同的需求。

规划布局

那不勒斯购物中心不是一个与外界鲜有互动的“密封盒”，而是一个与周边环境积极互动的媒体建筑，坐落在面积将近19 000 m^{2}的广场中心。

购物中心包括地上三层和地下两层，为不同的人群打造不同的商业空间，所以说它是一个多功能建筑。

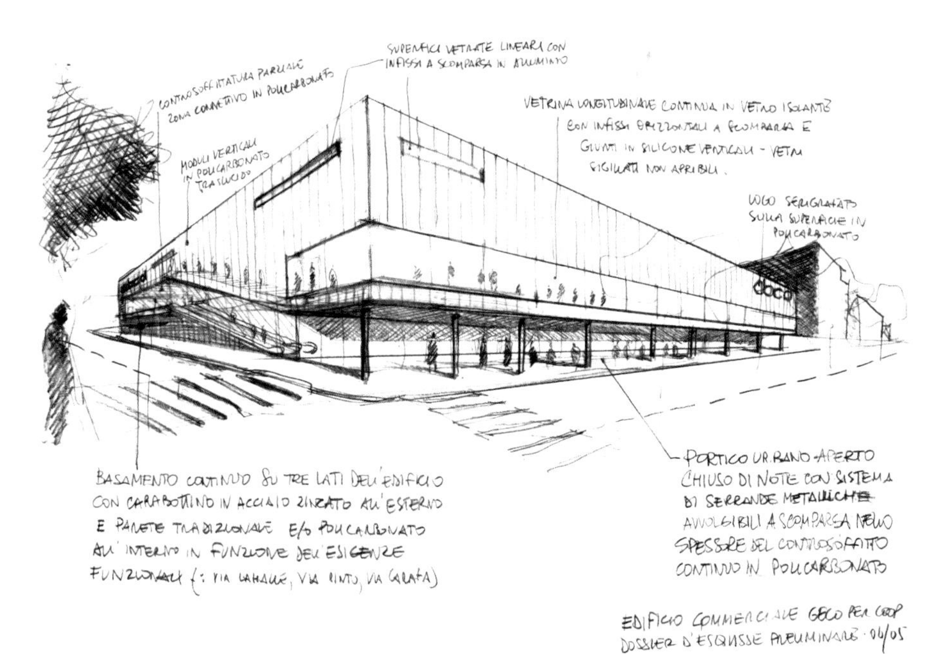

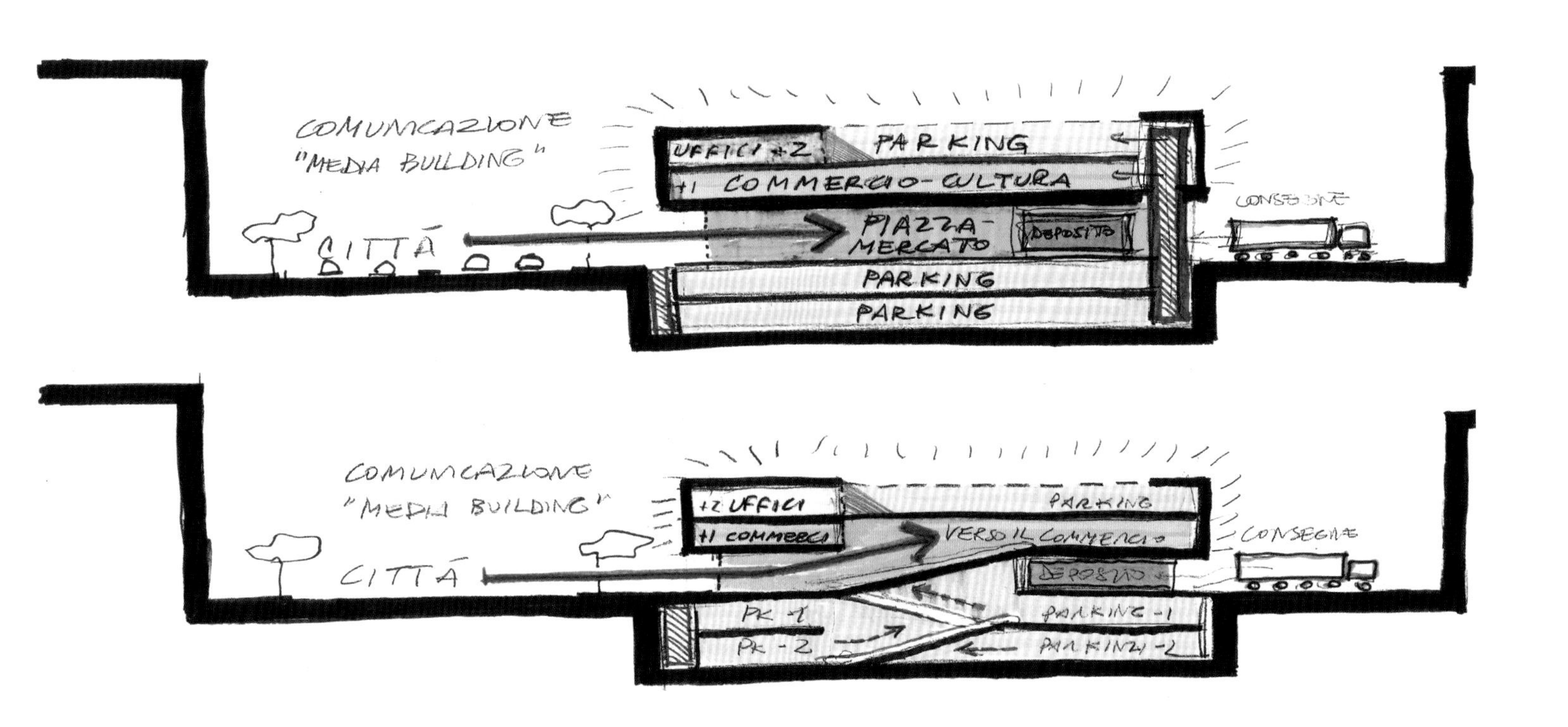

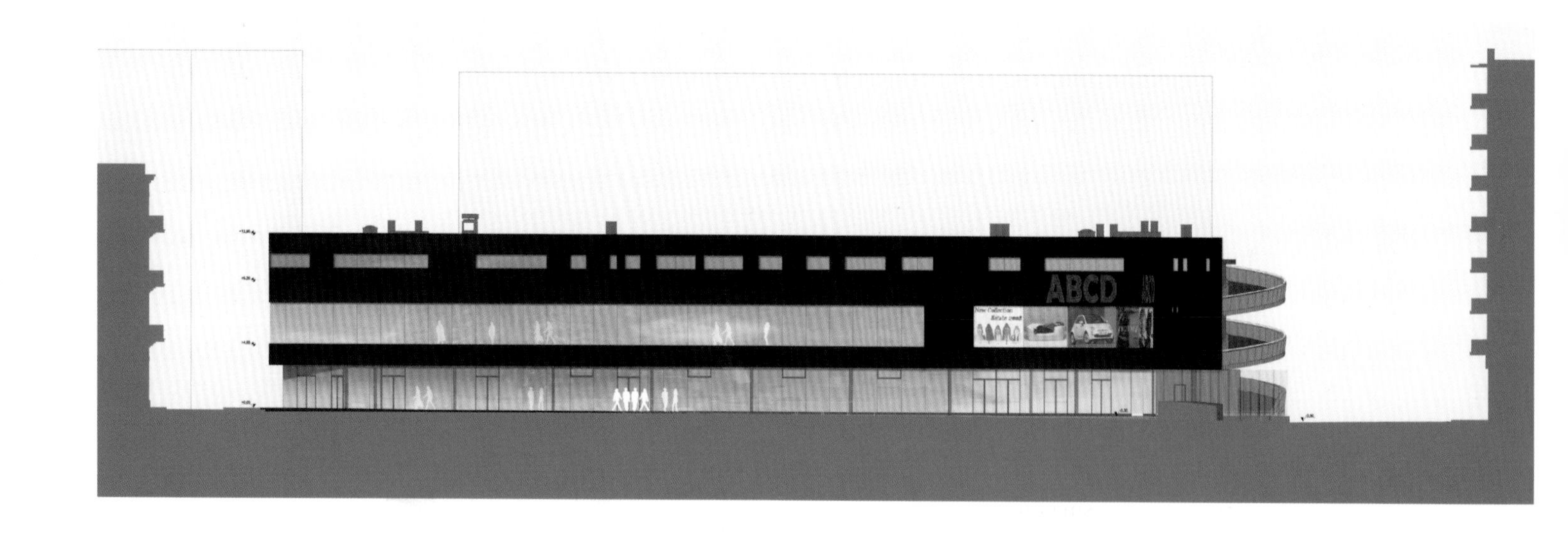

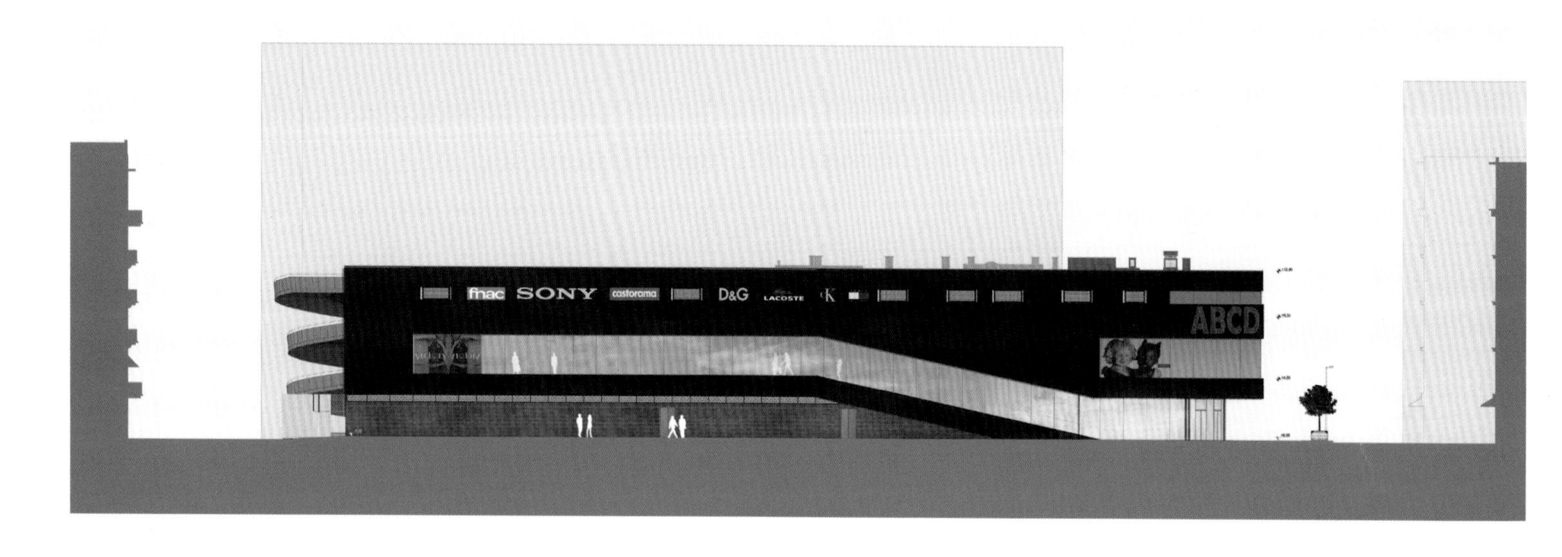

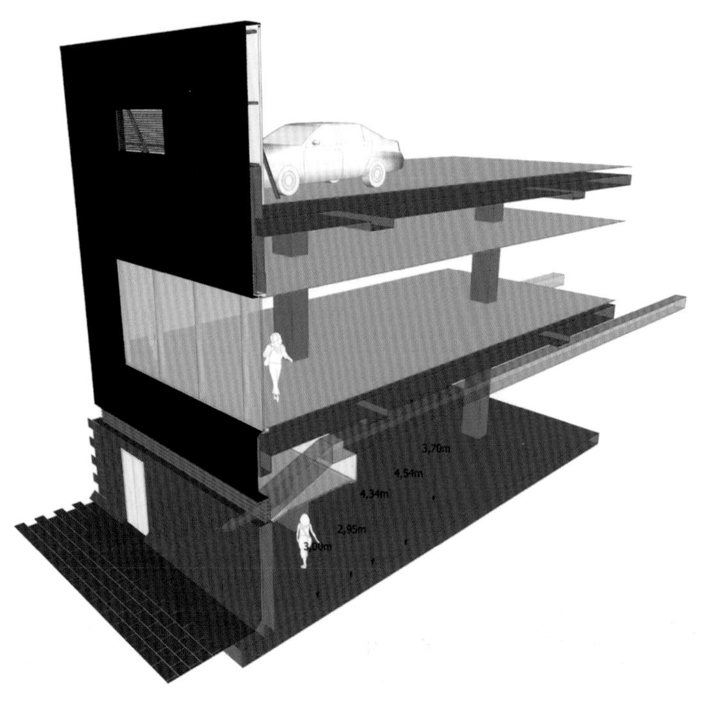

建筑设计

项目靠近Arenaccia大道的部分采用了大胆的建筑造型：一个中型的接待厅设置在地面层，商业、文化及媒体混合空间设置在二楼，从街上望去，透过横向玻璃，一目了然。地下层是停车区域，其30%作为商用，70%留给当地住户。

项目设计对周边环境产生的影响与外墙覆盖材料的选用密切相关。购物中心的外墙采用了一个极其简单的深色金属的模块系统，光泽的黑色面板体现城市环境，广告信息屏与黑色镶板形成鲜明对比，明显突出了媒体建筑的形象。

Profile

The transformation of Naple's former Post Office involves the project of a new kind of commercial complex in the Arenaccia housing estate and historical neighborhood. It is north of Naple's central train station going towards the airport. This residential part of the city is in need of leisure facilities, shops and of urban requalification in general.

Planning Layout

Instead of the usual airtight box in which commercial spaces are placed with no interaction with the exterior, the building is situated in the center of a square lot (approximately 19,000 m^2) and has been concieved as a media-building as it comunicates and transmits the functions of the use of spaces from the interior to the exterior.

The building is composed of 3 levels above ground and two below. Different commercial spaces are available to inhabitants of the area making this a mixed usage building.

Architectural Design

Along via Arenaccia the commercial spaces reveal themselves thanks to a bold architectural shape: on the ground floor there is a medium-sized reception hall

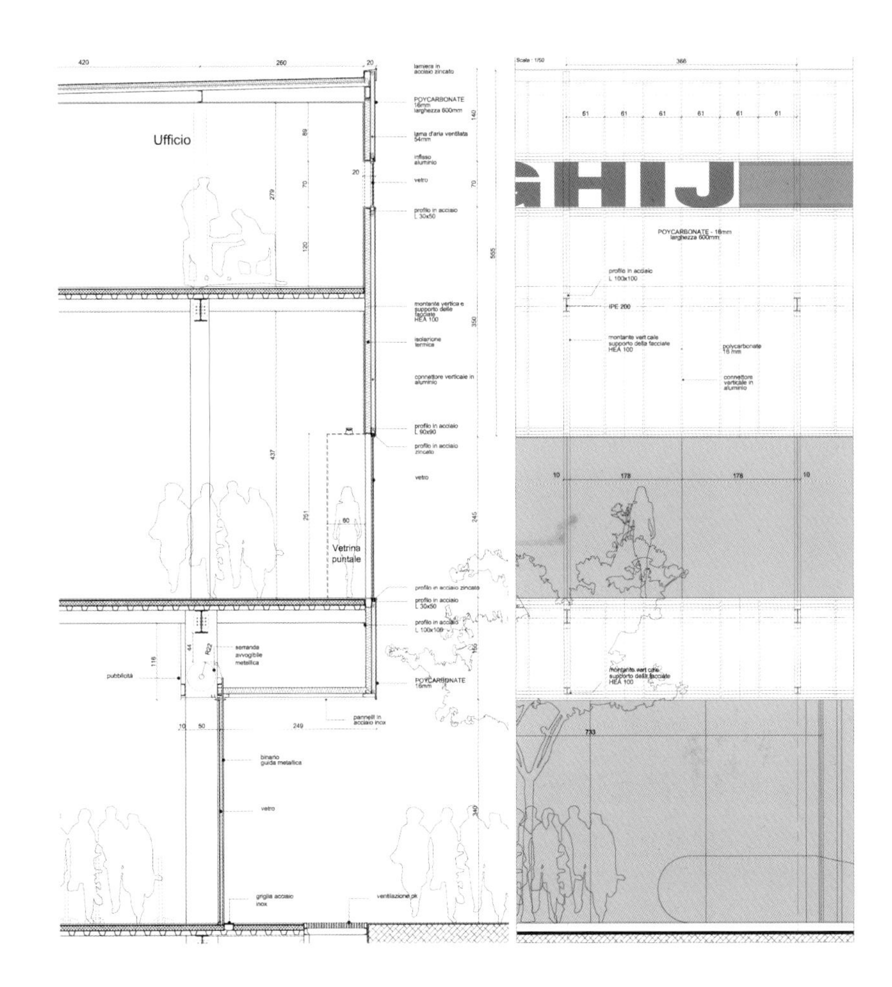

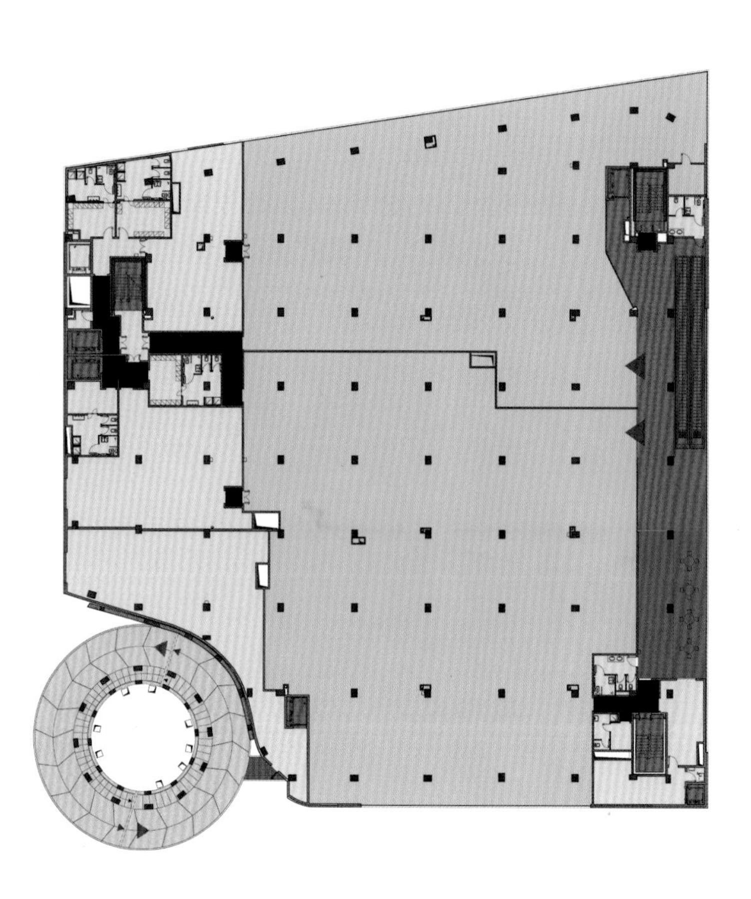

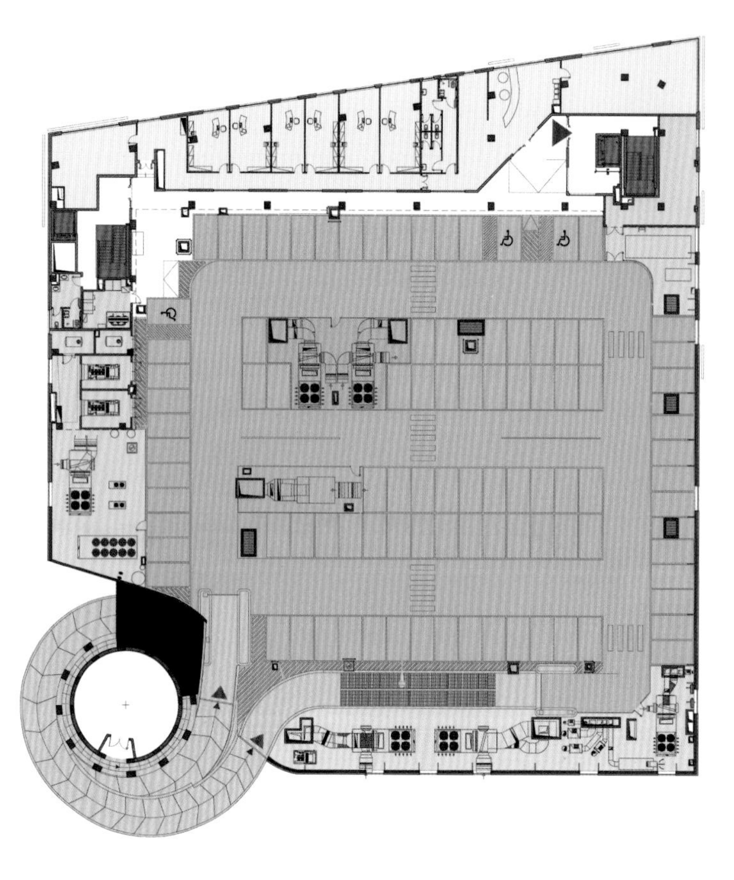

SATURN

opens onto the arcade, the first floor is characterized by the presence of commercial, cultural and mixed media spaces which can be seen from the street through a longitudinal window. The two floors below ground level are dedicated to parking spaces, 30% of these are for the above businesses while the other 70% is reserved parking for the residents of the area.

The impact on the surrounding area is tied to the choice of material used for the exterior cladding, using a very simple modular system of dark metal. The glossy black panels (erected on-site on a vertical 55cm wide grid) reflect the urban context and make the media building image evident, this is due to the contrast between publicity or information screens (comunicating surfaces) and the black paneling.

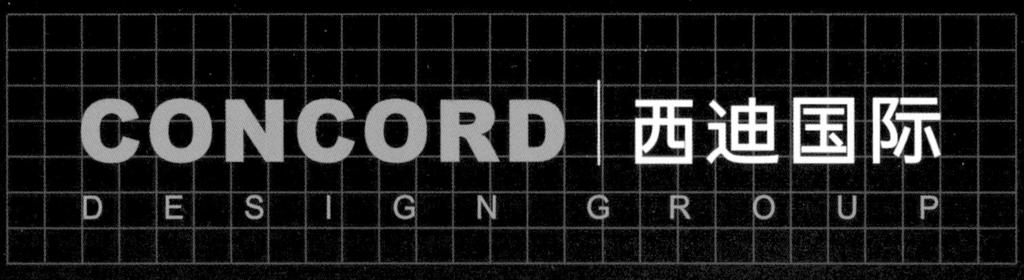
CONCORD 西迪国际
DESIGN GROUP